Contents

THE MOUNTAINS OF GREAT BRITAIN

Dane Love

© Dane Love, 2022
First Published in Great Britain, 2022.

ISBN – 978 1 911043 13 3

Published by Carn Publishing Ltd.,
Lochnoran House,
Auchinleck,
Ayrshire, KA18 3JW.

www.carnpublishing.com

Printed by Bell & Bain Ltd.,
Glasgow, G46 7UQ.

THE
MOUNTAINS
OF
GREAT BRITAIN

Scottish Kirkyards
The History of Auchinleck – Village and Parish
Pictorial History of Cumnock
Pictorial History of Ayr
Scottish Ghosts
The Auld Inns of Scotland
Guide to Scottish Castles
Tales of the Clan Chiefs
Scottish Covenanter Stories
Ayr Stories
Ayrshire Coast
Scottish Spectres
Ayrshire: Discovering a County
Ayr Past and Present
Lost Ayrshire
The River Ayr Way
Ayr – the Way We Were
The Man Who Sold Nelson's Column
Jacobite Stories
The History of Sorn – Village and Parish
Legendary Ayrshire
The Covenanter Encyclopaedia
A Look Back at Cumnock
A Look Back at Girvan
A Look Back at Ayrshire Farming
Ayr Then and Now
Ayrshire Then and Now
The History of Mauchline – Village and Parish
The Galloway Highlands
Ayrshire's Lost Villages
A Look Back at Dalmellington
Scotland's Lost Villages
A Look Back at Irvine
A Look Back at Stirling
The History of Ochiltree—Village and Parish

Introduction

Who it was that first decided to list mountains in tables has long-since been forgotten, and the tables they produced are no longer extant. Perhaps Montford John Byrde Baddely, the famous English fell guide, was one of the first, who is commemorated by a monument at Windermere in the Lake District. Regarded as 'the thorough guide', his tables of mountains in Scotland came to the grand total of thirty-one! He probably just made a tour round the highlands and jotted down those peaks he thought could be in excess of 3,000 feet. He certainly couldn't have asked the locals, for one is apt to overestimate the heights of mountains, as did the writers of the early *Statistical Accounts of Scotland*. The minister who wrote the description of the parish of Durisdeer, which occupies the western slopes of the Lowther Hills (Section 40) in Dumfriesshire, stated with confidence that Green Lowther, the highest of the Lowther Hills, was 3,130 feet hight, whereas in fact the Ordnance Survey have measured it to be just 2,403 feet tall.

But of course, Scotland was still a rather unexplored country, particularly to those from the south, who began to make expeditions north following the Union of Parliaments in 1707, just to see what this savage country was really like. Many of these travellers wrote of their experiences in books in order to educate the southerners, and which today make interesting, if amusing in places, reading. Daniel Defoe, of *Robinson Crusoe* fame, was one of the first who made this journey, in 1698, although he was a spy at the time, for he was having a good look at the land prior to the union being agreed upon.

Thomas Pennant, in his *Tour of Scotland* in 1769, wrote of the mountains:

Benevish [Ben Nevis] soars above the rest, and ends, I was told, in a point (at this time concealed in mist) whose height from the sea is said to be 1,450 yards.

As an antient Briton, I lament the disgrace of Snowdon; once esteemed to the highest hill in Britain, but now must yield the palm to a Caledonian mountain. But I have my doubts whether this might not be rivalled, or perhaps surpassed, by others in the same country; for example Ben y Bourd [Beinn a' Bhuird], a central hill, from whence to the sea there is a continued and rapid descent of seventy miles, as may be seen by the violent course of the Dee to Aberdeen. But their height has not yet been taken, which to be done fairly must be from the sea.

The Ordnance Survey, formed in 1791 to make an official map of Britain, has since surveyed the whole island, noting the heights of most summits, although the last levels were only taken in 1985, over two centuries after Pennant. Until that task had been completed, any list of mountains in Great Britain would have contained inaccuracies and estimates, due to the lack of detailed surveys of some uplands. During the nineteenth century the map-makers surveyed only some contour-lines, mainly the lower ones which would have had the greatest contribution to society, required, for example, for building and road alterations. The mountainous regions of Wales, England and the Southern Uplands of Scotland only had contour lines surveyed at irregular intervals, usually at significant heights. For example, in the Cumbrian mountains, contours were only surveyed at 50 feet, 100 feet, and every 100 feet thereafter to 1,000 feet, then every 250 feet to 2,000 feet, then every 500 feet to 3,000 feet. Those fifty feet contours which were shown on the map between these levels were the result of the map-makers' imagination, drawing them in just to show how he saw the shape of the land.

Things in the Scottish highlands were nowhere so 'good', for no contour lines were surveyed at all. There, only the odd spot height was marked on the map, the approximate levels of other peaks worked out from those summits which were measured. The cartographer then drew in the relevant number of contour lines to give the hills their approximate shapes. This was not even done from 'on-site' skills, but done back in the Ordnance Survey offices, using as a guide the old maps of the country where hills were only shown by hachures.

Because of this, contour lines could indicate that a hill was a shape that it wasn't in real life. In other cases the height of the hill could be wildly inaccurate, no summit levels known, and the map-makers guessing its height from the comfort of their English office! An example of this is Creag Doire na Nathrach in the Glenlyon and Mamlorn section. According to the One Inch map the hill was between 2,250 and 2,300 feet high (685-701 metres), according to the fifty feet contours shown on the map. However, following the new survey of the whole country at ten metre intervals, the mountain is now found to be 837 metres in height, over 140 metres taller than first indicated!

Today, on maps of the highlands, the contours are shown at ten metre intervals and, though there have still been few proper levels taken, these contours are far more accurate than previously shown. The contour levels were worked out by aerial surveys, combined with a process known as analytical aerial triangulation, and then plotted on the maps.

The heights of the summits over 3,000 feet were better known, for Sir Hugh Munro measured these with an aneroid barometer when he made his tables. Sir Hugh Thomas Munro, Baronet of Lindertis, was a founder member of the Scottish Mountaineering Club in 1889. Munro's Tables are the best known set of mountain tables in Britain, and date from 1891 when they were first published in the Scottish Mountaineering Club Journal. Sir Hugh Munro climbed every summit bar two, Carn a' Cloich-mhuilinn in the Monadh Ruadh, and the Inaccessible Pinnacle on Skye, which he had been keeping for last. However he died in 1919 of Spanish Influenza before he was able to reach these summits.

The tables he left us have remained popular ever since, and in fact are more popular than ever. The Scottish Mountaineering Club has kept them up to date, making changes as new surveys alter the heights, and adding those peaks now found to be in excess of 3,000 feet but which Munro regarded as below. Only Beinn an Lochain in Argyll did Munro include which he should not have. As stated, Munro didn't climb two peaks on his list, the first person to do so being A. E. Robertson, reaching his last summit cairn in 1901. The Munro tops (lesser summits) were first climbed in 1923, and the Munros in the rest of the British Isles, or 'furth of Scotland' as they are known, in 1919. John Dow was the first to do the three groups. The SMC notes all who climb the Munros, up to two hundred or so new walkers every year. There are currently 282 summits over 3,000 feet that are listed as Munros.

Following the publication of Munro's Tables, other tables followed, the compilers inspired by his example. J. Rooke Corbett, the fourth Munroist, decided to compile the Welsh and English 'Munros'. However, there were relatively few (around 20), so he altered his criterion to summits over 2,500 feet in height. His tables were published in the Rucksack Club Journal in 1911, listing 131 tops, but a 1929 revision increased this figure to 148.

Corbett then went further, deciding to list the Scottish 2,500 feet summits. Because of the large number of hills which would then qualify for inclusion, he changed the criteria again. To be included in the Scottish Corbetts, a hill must have a re-ascent of 500 feet. Although this excluded many tops, the list did extend to 217 peaks. Corbett did not publish this list, for he probably wished to make a further revision, but he died before he was able to do so. His manuscript was passed on to the SMC who included it in later copies of Munro's Tables. A number of errors have since been noted, no doubt Corbett would have discovered them himself, but he did enjoy the distinction of being the first to climb all summits in his own list. There are now 222 Corbetts in Scotland listed as meeting the specifications laid down by Corbett himself.

A list of the summits over 2,000 feet in England and Wales was then published by Rev W. T. Elmslie in the Fell & Rock Climbing Club Journal in 1933. Elmslie also listed every spot height shown on the maps, so that every 'hill' on his list need not have a summit, as it could simply be a point on the side of a hill. The maps he used were Bartholomew's Half Inch series, the contours 250 feet apart, so that his list was not particularly accurate or consistent, although it did include 347 points, of which only 305 are actually summits. The first person to compile the English and Welsh 2,000 footers with true summits was E. Moss, whose lists of 1939 (England) and 1940 (Wales) had 612 mountains, later revised to 621.

George Bridge compiled his own list of *The Mountains of England and Wales* in 1973 which became one of the most accepted lists for climbers in the two countries. It extended to 408 tops. To qualify, a mountain had to have a re-ascent of 50 feet, resulting in many lesser knolls being included. A similar list was published in 1986 by Chris Buxton and Gwyn Lewis, whose *The Mountain Summits of England and Wales* (with a ten metre re-ascent) extends to 423 tops.

A list of 2,000 foot summits in the Southern Uplands of Scotland had already been compiled by Percy W. Donald. This list has 133 summits. In Ireland the mountains were listed by C. R. P. Vandeleur and Joss Lynam, but have become known as the 'Wall's', because these tables were first published in *Mountaineering in Ireland* by Claud Wall (1976).

The first compilation of the 'Loves', as the present listing was known by some, took place in 1983, and these were almost published on two occasions, but financial restrictions meant that they did not see the light of day. An updated version of this listing is what this book contains.

Other list compilers are worth mentioning in

passing. F. H. F. Simpson produced lists of mountains in England in 1937, and in 1939 made improvements to this. In 1940 he included Wales for the first time. W. MacKnight Docharty produced an improved list of English mountains over 2,500 feet, and in 1962 issued a new list of mountains in England and Wales over 2,000 feet. In 1974 Nick Wright compiled a list of *English Mountain Summits*. From 1989 onwards John and Anne Nuttall published the *Mountains of England and Wales*, having relisted all of the summits according to the criteria of being in excess of 2,000 feet and with a re-ascent of at least 50 feet. They did not produce tables as such, but detailed guides on each summit and routes to ascend them. In total, they reckon that there were 441 summits in total, 253 in England and 188 in Wales.

A further list of mountains over 2,000 feet in England and Wales with a 100 feet re-ascent was compiled by Alan Dawson. Totalling 525 summits, they have become known as Hewitts, from 'Hills in England, Wales and Ireland over Two Thousand feet'.

In Scotland, Fiona Graham, later Torbet, compiled lists of hills in Scotland over 2,000 feet, but less than 2,500 feet, with a re-ascent of 490 feet (150 metres), or basically Corbett criteria. Alan Dawson also compiled a list using the same criteria at much the same time, totalling 219, and in 2004 extended the definition downwards to summits with a prominence of 30 metres, naming these as Graham Tops, totalling 776.

The Database of British and Irish Hills was created in 2001 and lists over twenty thousand summits across the British Isles, from county tops, to TuMPS, hills (of any height) with a re-ascent of 30 metres.

Each compiler has used different criterion in their definition of a 'mountain', and arguments continue among climbers to this day just what constitutes a mountain, and what is a hill. Some folk claim that a summit must be over 3,000 feet to be a mountain, others that it is over 2,000 feet. I know of a geography teacher who insisted that in Britain the height which separated mountains and hills was 2,300 feet, so that my nearest mountain (Blackcraig Hill - according to my list) was just a hill, failing to reach the magical figure by 2 feet.

Walt Unsworth, in his *Encyclopaedia of Mountaineering* states that a mountain in Scotland is a hill over 3,000 feet tall, in England over 2,000 feet. This makes a mockery of Windy Gyle, the Cheviot summit on the border, one side of which must be a mountain, the Scottish side merely a hill! Some people go so far as to claim that there are no mountains in Britain, our summits just being hills when you compare them to the Alps or Himalayan peaks. A friend defines a mountain as being rugged, compared to the smooth lines of a hill, whereas the dictionary I normally use simply, if blandly, defines a mountain as 'a high hill'.

Since originally compiling these lists, the term 'prominence' has become a more common means of identifying mountains and summits of importance. Prominence, or autonomous height, refers to what others called 're-ascent', that is, the height of a summit down to the lowest point on the col or pass between it and a taller, or 'parent' peak. Using this definition, Scafell Pike has, for example, a summit height of 978 metres, a prominence of 912 metres (the next highest summit to which it is attached being Snowdon), but the adjoining Sca Fell with a height of 964 metres only has a prominence of 133 metres, the Mickledore pass having a height of 831 metres.

It is interesting to note that all definitions of mountains in Britain use the imperial system of measurement, no-one claiming mountains to be in excess of 700 metres, say. Similarly on the continent, where the Metric system of measurement has long been established, a mountain will hardly be defined in feet. Now that Britain has been metricated, at least the maps have, if we still walk for miles rather than kilometres, will future generations claim their mountains are over 609.6 metres, or conveniently 'metricate' our mountains to 600 metres?

These tables have plumped for the 2,000 feet contour as the separating border between hill and mountain, mainly because 3,000 feet is rather high (and Munro has listed them already) and because 2,000 feet has been adopted by most compilers in the past. By using 2,000 feet many summits in the Highlands of Scotland are included which have never appeared in tables before. This is what makes these tables unique, and the fact that they apply the same criteria to the whole of Great Britain. No doubt some day someone will produce a list of mountains in the British Isles - I didn't because Ireland is not covered by the Ordnance Survey, has a different National Grid and base level from which heights are measured.

Although this book lists mountains over 2,000 feet, the heights are in metres. The reason for this is that the Ordnance Survey maps are now metricated, ten metres being the vertical interval of the contours, and the heights listed in metres. To convert heights back into feet would not work, for three feet make up a metre, and which height to choose in the conversion couldn't be made without reference to Ordnance Survey files which list some heights to the nearest 100mm. The old One Inch Map (with heights in feet) could not be used, for the whole country has since been resurveyed, and many heights have since changed, more accurate surveys giving better defined heights.

A second criterion has also been used by many compilers in making their tables of mountains. That is the height of re-ascent which separates mountains and mere 'tops' or knolls which are part of a greater height. Corbett in his Scottish tables makes his re-ascent 500 feet. Bridge in his tables of England and Wales uses a rather complicated system in which distance separating summits is also noted along with the re-ascent. Both figures are applied to a graph which indicates whether a summit is a mountain or just a top. Some compilers do not separate mountains and tops at all, listing every hill which is identifiable on the map as having a summit of some sort, the re-ascent therefore being as little as two metres in some cases.

Which of these criteria to follow in these tables was difficult. 500 feet seemed too much, Bridge's system too complicated, walkers on the hill being unable to tell from the map whether a peak was listed or not. In the end I have opted for 100 feet, or 30 metres, which seems to work well in most areas, although in some rugged parts of the highlands little peaks on ridges merit inclusion. Where there is doubt of a hundred feet of re-ascent I have opted for the safety of inclusion within these tables.

With metrication having established itself in Britain since this list was first compiled, I hope that this book will go some way to keep the mountains of Great Britain a traditional group, based on our ancient mensuration system, and adopted as the standard list. The climbing of Munros remains the most popular use of tables, so for those who like to count them I have indicated them (and also Corbetts) in the lists.

Those living far from the highlands, yet still enjoying hill-walking, can tick off those summits nearer to them which they have climbed. There being in excess of 2,500 mountains listed (almost nine times the number of Munros) climbing them all will be a very great challenge, and whether anyone ever manages to do so remains to be seen. Unlike Corbett, I doubt very much whether I will ever be the first to climb my own list!

Although comprising hill summits that do not meet the specification for inclusion in this book, the existence of the Hughs should be noted. These are 'Hills Under Graham Height', the initial letters forming the name. These hills are deemed as being of significant 'attitude' if not altitude, and include many smaller hills that are prominent in their own locality, either from their shape or ruggedness.

Another group of mountains have been named 'Archies' and these are the summits in Scotland that are in excess of 1,000 metres with a re-ascent of 100 metres. These total 130 and were first listed in 2015.

There are a handful of hills whose inclusion in this list are subject to some doubt. There are a few summits which appear to be almost 2,000 feet in height, and depending on who measures them, and what datum is used, results in different figures over the years. One of these is fairly local to me—Dugland in the Afton and Carsphairn hills—which was originally noted by the Ordnance Survey as being 2,000 feet in height. Consequently it was included by Percy Donald in his tables, and was accordingly climbed by myself in the completion of his list. Later surveys, however, noted the hill as being just too low, and therefore I had to remove it from an intermediate version of my list. Latest maps show the summit as being 612 metres above sea level, so the hill has been promoted once again.

Some other summits across the country have similar tales—in some cases resulting in them being added to the list of 2,000 footers. Mynydd Graig Goch in North Gwynedd has been confirmed as reaching the 2,000 foot mark. Thack Moor in the North Pennines is confirmed as being 2,000 feet in height. Similarly, Calf Top in the Southern Pennines has been accurately measured and its height confirmed as over 2,000 feet by just 6 millimetres. Many of these precise surveys were carried out by a trio of keen mountain surveyors, whose definitive heights have been accepted by the Ordnance Survey.

Explanation of Tables

The country is divided into 59 districts, ranging from the north of Scotland southwards to Dartmoor, finishing with the Isle of Man. At the start of each district section is an introduction to the area, detailing some of the more interesting mountains within it, in some cases popular routes, and something of the history or ownership of the countryside.

The tables list mountains with the height in metres to start with. These heights are sourced from Ordnance Survey maps, usually at 1:25,000 scale. In a few cases, heights have been obtained from 1:50,000 mapping (where the larger scale map doesn't include a spot-height). In a random few cases, heights have been obtained from other sources. Where efforts to find a definitive height for a mountain have failed, an estimated height is given, and this is indicated by 'est' appearing after the height.

To the left of the height is a square box which hill baggers may wish to use to indicate that they have ascended that summit. The box can be ticked, crossed or coloured as the user wishes.

The name of the mountain follows in the next column. In a number of cases there is more than one name given. This can be for a variety of reasons, such as English/Gaelic or Welsh/English versions of the same name, or where there are alternative names for some summits (it's surprising how many mountains have different names that have been adopted by various writers and guidebooks over the years). In a few occasions, names have been devised for some summits, particularly in remote corners of Scotland, where no name appears to exist for the summit. The creation of new names is not unique to this list, George Bridge listed summits with 'Author's temporary name' in the notes. In many cases these names have since been adopted.

In some cases the mountain name column has a 'range' or grouping title, written in italics. This is to indicate a distinct group of mountains that have a known locality name. The mountains that fall within this locality are indented in the subsequent list.

At the start of some mountain names are symbols, indicating that the peak also qualifies as a mountain as indicated in the following lists:

▲   Munro (Scottish summit over 3,000 feet)
△   Corbett (Scottish summit over 2,500 feet)
●   Munro Furth of Scotland
⊙   Corbett Furth of Scotland

More details about what qualifies for the above definition is given in the Introduction.

The next column gives the Ordnance Survey National Grid Reference for the summit of the mountain. This includes the Grid Letters plus a six-figure reference.

The next two columns lists the Ordnance Survey maps on which the mountain summit appears. The first column is the numbers of the Landranger (1:50,000) maps. The second column lists the 1:25,000 maps sheets that the mountain appears on. In some cases they are referred to as Explorer maps, in other cases as Outdoor Leisure maps. However, the numbers do not overlap.

The final column contains six squares. These can be used to write in the date of ascent of the particular mountain, or if one climbs mountains more than once, second and subsequent ascents can be marked off.

Reay

The area covered by Reay in this book is the wild countryside of Sutherland north of the Loch Shin, Loch More, Loch Stack glen (traversed by the A838), and west of Loch Loyal and the A836. The Reay Forest is an ancient deer forest, but the name historically extended far wider over much of north-west Sutherland.

There are many distinctive mountains in this area, though only one, Ben Hope, attains Munro status. To the west are the great summits of Arcuil (Arkle) and Foinne Bheinn (Foinaven) rising from a wide moor spattered with lochs. With Creagan Meall Horn, these mountains form part of the North-west Sutherland Area of Outstanding Natural Beauty. Many of the mountains are composed of quartzite, layered over the gneiss of the moors. Foinne Bheinn was at one time thought to be a Munro but an accurate survey confirmed that it failed to reach the magic height by twelve feet. The view over the loch-spattered gneiss moors towards Loch Laxford is one of the finest in this part of the country. On the east side of Foinne Bheinn are a series of corries, overlooking Srath Dionard. The two summits of Arcuil guard the hidden corrie of Am Báthaich. By studying Arcuil and Foinne Bheinn, nineteenth-century geologists were able to understand how the mountains of the Alps and Himalaya were formed.

South of here are the heights of Meallan Liath Coire Mhic Dhughaill, Cárn an Tionail and Beinn Direach. The rounded mountains have broken rocks on them in various places, and some attractive corries, in many cases with lochans, and the lower height attained means that they are quieter than the usual Munro summits. Only Meallan Liath Coire Mhic Dhughaill is visited more than the usual, being a Corbett. Much of this countryside is owned by the Duke of Westminster as a sporting estate.

Across the Bealach nam Meirleach rises another

1 Corbett, Beinn or Ben Hee, from the Gaelic Beinn Sith, or fairy hill. It is covered in loose rocks and large boulders, and from the summit, when it is clear, it is possible to see Beinn Laoghal and Cuinneag.

A separate area of outstanding landscape beauty is the Kyle of Tongue district, between lochs Hope and Loyal, wherein rise the majestic peaks of Beinn Laoghal and the prominent prow of Ben Hope. Beinn Laoghal is a granite mountain, the only one this far north, with a series of jagged peaks, often appearing in calendars. However, the 100 feet of re-ascent definition used in this book results in the mountain being divided into five tops. An Caisteal is the highest, but to the north of this is Sgór Chaonasaid, the northern slopes of which are formed of large slabs of rock. Beinn Bheag is sometimes referred to as Heddle's Top. Cárn an Tionail is the least-interesting of the Laoghal tops, a rounded summit at the southern end of the ridge. West from Beinn Bheag is Sgór a' Chleirich, a narrow ridge with the highest point at the top of a rock face overlooking Coire Fhionnaich.

Ben Hope has the distinction of being the most northerly Munro, and as it rises almost on its own, forms a major rock mountain when viewed from the north. The most common ascent starts from Muiseal at the Strathmore River, a steady climb over the southern flanks being straightforward. The east side of the mountain has a series of corries, and the north ridge provides a more airy and in some parts difficult means of ascending the mountain for more experienced climbers.

Two further mountains rise west of Loch Eriboll and south of the Kyle of Durness. Beinn Spionnaidh and Crann Stacaidh are sufficiently high and separate to qualify as Corbetts in their own right, but are usually ascended together. The mountains comprise of broken quartzite which can make for difficult walking. Ascents from the west are probably more interesting than from the east, where the hills are simply steady slopes from sea level.

The Mountains of Great Britain

Height	Name	NGR	OS L	OS E	Ascent
773	△ Beinn Spionnaidh	NC 363573	9	E446	
801	△ Crann Stacaidh (Cranstackie)	NC 351555	9	E446	
	Foinne Bheinn:				
902	Ceann Garbh	NC 313514	9	E445	
911	△ Ganu Mór	NC 317507	9	E445	
869	Foinne Bheinn (Foinaven)	NC 319494	9	E445	
808	Stob Cadha na Beucaich	NC 325487	9	E445	
778	An t-Sàil Mhór	NC 339484	9	E445	
787	△ Arcuil (Arkle)	NC 303464	9	E445	
758	Arcuil Deas	NC 310453	9	E445	
731	Creagan Meall Horn	NC 345457	9	E445	
777	△ Meall Horn	NC 353449	9	E445	
732	Sàbhal Beag	NC 374429	9	E445	
720	Beinn Stac	NC 269423	9	E445	
718	Sáil Rac (Tátha nam Beann)	NC 356403	9	E445	
754	Meall Garbh	NC 369404	9	E445	

1

	Height	Name	NGR	OS L	OS E	Ascent
☐	761	Stob Coire nan Loch	NC 361398	16	E445	
		Reay:				
☐	801	△Meallan Liath Coire Mhic Dhughaill	NC 357392	15	E445	
☐	797	Cárn Dearg	NC 378389	16	E445	
☐	759	Cárn an Tionail	NC 392390	16	E447	
☐	752 est	A' Ghlaise	NC 391377	16	E447	
☐	688	Beinn Direach	NC 407380	16	E447	
☐	631	Meall a Chléirich	NC 408365	16	E447	
☐	683	Meallan Liath Mór	NC 406329	16	E440/E447	
☐	873	△Beinn Hee	NC 426339	16	E440/E447	
☐	851	Stob Coire na Sigh Duibhe	NC 434350	16	E447	
☐	927	▲Ben Hope (Beinn Hop)	NC 477501	9	E447	
☐	718	Sáil Romascaig	NC 492482	9	E447	
		Beinn Laoghal:				
☐	712	Sgór Chaonasaid	NC 579499	10	E447	
☐	765	△Beinn Laoghal (An Caisteal)	NC 578489	10	E447	

Beinn Bheag (Heddle's Top)   NC 577483   10   E447
Sgór a' Chléirich   NC 568485   10   E447
Cárn an Tionail   NC 575477   10   E447

☐ 744
☐ 663
☐ 716

Farr & Clyne

These mountains are located to either side of Loch Choire, and Loch Choire Forest. On the north-west side of the glen rises Beinn Cleith Bric, commonly named Ben Klibreck on many maps. The highest point of this mountain is found at Meall nan Con, which qualifies as a Munro, the second northernmost in the country. Two subsidiary tops qualify as mountains in this list, one to either side of the highest summit—Creag an Lochain to the south-west and Meall Ailein to the north-east. To the south-east rises Meall an Eóin, which just fails to qualify as it only has 29m of re-ascent. Beinn Cleith Bric is usually ascended by a long walk from Vagastie Bridge to the south west.

On the south-east side of Loch Choire are a series of hills that gradually increase in importance to the east, where the summit of Beinn an Armuinn is found. Also spelled Ben Armine, this summit is not the highest in this group, but it gives its name to the Ben Armine deer forest, a property of Sutherland estates. Its rocky eastern slopes and prominence make it more distinctive than its southern, and slightly higher, neighbour, Creag Mhór. This is a fairly rounded summit, only on its steeper eastern slopes having rocks break through the surface. With a couple of other summits, Meall nan Aighean and Creag na h-Iolaire, these mountains make a long day's hill-walking, a fair distance needing to be traversed before the foot of the range is reached. This is often from the public road at Strath Brora, via Ben Armine Lodge, or else from the road at Badanloch Lodge, heading westwards towards Loch Choire Lodge. Creag na h-Iolaire just qualifies as a separate mountain, there being only 108 feet or so of re-ascent from Meall nan Aighean.

The most attractive part of this range is within the Loch Choire glen, from where the hills rise steeply above wooded lochsides—Loch a' Bhealaich and Loch Choire itself. There is only two feet of difference in the level of water in these lochs.

Height	Name	NGR	OS L	OS E	Ascent
	Beinn Cleith Bric:				
☐ 808	Creag an Lochain	NC 576280	16	E443	
☐ 962	▲Beinn Cleith Bric (Meall nan Con)	NC 585299	16	E443	
☐ 724	Meall Ailein	NC 613315	16	E443/E448	
☐ 694	Creag na h-Iolaire	NC 673289	16	E443	
☐ 695	Meall nan Aighean	NC 681289	16	E443	
☐ 705	Beinn an Armuinn (Ben Armine)	NC 695273	16	E443	
☐ 713	Creag Mhór	NC 698240	16	E443	

25

Langwell & Loth

These four mountains are located on the eastern part of Sutherland, where the landscape tends to comprise of wild moors, sometimes peat flows, covered with a plethora of lochans. This part of the Highlands is rather bare of summits over 2,000 feet, however the four summits that qualify for this list are decent enough hills in their own right.

The easternmost summit is Scaraben, which forms a long ridge of summits almost on an east-west axis. Access to the summit requires a long walk-in over open moorland, either from Braemore to the north, or from the track that follows the Langwell Water. West of Scaraben is Srón Gharbh, a summit that attains 609m, and thus almost qualifies for inclusion in the list.

The same two access roads can be used to reach Mór Bheinn, the highest mountain in this part of the country. It forms a prominent conical hill, rising quite a bit more than the adjacent hills, and much more than the surrounding moors. The hill has screes and rocks round all sides, interspersed with heather slopes. Like the nearby Maiden Pap, it has its own 'nipple'.

The remaining two hills are located to the west of Glen Loth, further south, west of Helmsdale. A minor road that links the coast road at Lothbeg to the Strath of Kildonan passes along the eastern flanks of the two summits, the slopes on this side being rather rocky. A third, lower summit, Druim Dearg, forms the trio of summits seen from this side. Beinn Uarie is the northmost of the two mountains, the top indicated by a trig point in addition to a prehistoric cairn. Ancient tradition states that Clach Mhic Mhios, a standing stone on the moor to the east, was thrown from the summit of Beinn Uarie by a youthful giant. Beinn Dhorain forms the southern half of what is really one large massif, and it rises five metres more in height. Both summits are easily ascended from the minor road at the head of Glen Loth, a short round being possible within an hour or so.

3

Height	Name	NGR	OS L	OS E	Ascent
☐ 626	Scaraben	ND 067268	17	E444	
☐ 706	Mór Bheinn (Morven)	ND 004286	17	E444	
☐ 623	Beinn Uarie	NC 928164	17	E444	
☐ 628	Beinn Dhorain	NC 925156	17	E444	

Assynt & Coigeach

Some of Britain's finest mountains are located within this section—peaks such as Stac Pollaidh, Suilbheinn, Canisp, Cuinneag and the two Culs, Beag and Mór. They rise like huge sentinels of Torridonian sandstone above the Lewisian gneiss rock moors, a landscape covered with numerous oddly-shaped lochans amid the heather. With steep rocky sides, the mountains make excellent scrambling, and even lower summits such as Stac Pollaidh, which just reaches over the 2,000 foot mark, give wonderful experiences.

At the north-east end of the section are a number of remote summits, rarely climbed as they fail to match the surrounding Munros for height or grandeur, but which individually form interesting walks. Beinn Leóid is a remote Corbett requiring long walks-in to reach. Most ascents are probably from Coire Ceann Loch to the east. East of this are the summits of Meallan a' Chuail and Meall an Fheur Loch, the latter just over 2,000 feet.

Cuinneag, or Quinag, is a rather fine mountain, and although it lacks Munro criteria, it has plenty of status of its own. In fact, the three main summits are Corbetts in their own right, and the lesser tops on the ridge have character of their own. The mountains have a layer of Cambrian quartzite over the sandstone.

To the east of the A894 rises another Corbett, Glas Bheinn, with a long ridge to Beinn Uidhe. South-east of here the countryside rises to the Munros of Con a' Mheall and Beinn Mór Assynt, the latter just a couple of metres short of 1,000 metres. Breabag north and south tops are attractive mountains, and the remote Meall an Aonaich, with the Eagle's Rock on the summit, makes an interesting expedition from Glen Cassley.

In Assynt rise the great monolithic mountains, Canisp to the north. It is a Corbett, the usual route of ascent being from the A837 to the east. From the summit the mile-long ridge of Suilbheinn is seen due west, one of Scotland's finest peaks. The western end (Caisteal Liath) appears like a sugar-loaf, whereas the east (Meall Meadhonach) is pointed. Most ascents are made from Inbhir Chircaig to the west, the initial walk along the Abhain Chircaig to the Falls of Kirkaig being attractive in itself.

Cúl Mór is a Corbett rising from the moine, but its bulk is greater than the adjoining mountains. Ascents are often made from a parking area at Lochan an Ais, where the Knockan Crag National Nature Reserve is located. A subsidiary top is Creag nan Calman, and the summit of Bod a' Mhadail almost merits inclusion as a mountain, the re-ascent being close to 100 feet.

To the south is Cúl Beag, the little sister of Cúl Mór. Ascents are usually made from the east end of Loch Lurgainn, allowing the cliffs to be avoided. Meall Dearg is a subsidiary top to Cúl Beag, but merits inclusion as a mountain. The extensive estate of Assynt, from Loch Assynt south to Cúl Beag, is owned by the Assynt Foundation, a community buy-out of a sporting estate.

Rising steeply at the west end of Loch Lurgainn is Stac Pollaidh, one of the district's finest peaks. A popular path ascends the slopes from a parking area at the roadside, and the ridge requires care when traversing.

From Stac Pollaidh one can look over Loch Lurgainn to Coigach, where a series of interesting peaks occupy a stretch of countryside between the loch and the sea. Sgórr Deas of Beinn an Eóin may only just merit inclusion as a mountain, but it is an interesting peak. Across Lochan Tuath is the tooth of Sgúrr an Fhídhleir and the elongated bulk of Beinn Mhór na Cóigich. With no local roads, this mountain and a few outlying tops provide remote walking. The countryside around here is owned by the Scottish Wildlife Trust. All of the mountains in this section are located within the Assynt-Coigach National Scenic Area, such is the importance of the landscape and nature therein.

4

Height	Name	NGR	OS L	OS E	Ascent
613	Meall an Fheur Loch	NC 362311	16	E442/E445	
750	Meallan a' Chuail	NC 344293	15	E440/E442	
792	△ Beinn Leoid	NC 320295	15	E442	
652	Sáil na Slataich	NC 314304	15	E442	
	Cuinneag:				
776	△ Sáil Ghorm	NC 194304	15	E442	
745	Cuinneag Meadhon	NC 201289	15	E442	
809	△ Cuinneag - Sáil Gharbh (Quinag)	NC 209292	15	E442	
714	Creag na h-Iolaire Ard	NC 202282	15	E442	
764	△ Spidean Cóinich	NC 206277	15	E442	
776	△ Glas Bheinn	NC 255265	15	E442	
741	Beinn Uidhe	NC 282252	15	E442	
640	Cadha na Poite	NC 289235	15	E442	
860	Na Tuadhan	NC 304215	15	E442	
987	▲ Con a' Mheall (Conival)	NC 303199	15	E442	
998	▲ Beinn Mor Assynt	NC 318202	15	E442	

4

	Height	Name	Grid Ref	Map	Sheet
☐	960	Cnap Ghiubhain	NC 324193	15	E442
☐	868	Càrn nan Conbhairean	NC 325183	15	E442
☐	715	Meall an Aonaich	NC 336164	15	E440/E442
☐	718	Breabag Tuath	NC 292180	15	E442
☐	815	△ Breabag	NC 287158	15	E442
☐	688	Meall a' Bhraghaid	NC 298142	15	E442
☐	847	△ Canisp	NC 203187	15	E442
☐	731	Suilbheinn - Caisteal Liath	NC 153184	15	E442
☐	691	Suilbheinn - Meall Mor	NC 162179	15	E442
☐	723	Suilbheann - Meall Meadhonach	NC 164178	15	E442
☐	657 est	Suilbheann - Meall Beag	NC 166177	15	E442

Inbhirpollaidh:

	Height	Name	Grid Ref	Map	Sheet
☐	849	△ Cul Mor	NC 162119	15	E439
☐	828	Creag nan Calman	NC 159113	15	E439
☐	769	△ Cul Beag	NC 140088	15	E439
☐	657	Meall Dearg	NC 148187	15	E439
☐	612	Stac Pollaidh (Stac an Poile)	NC 108106	15	E439

4

Height	Name	NGR	OS L	OS E	Ascent
	Coigeach:				
619	Beinn an Eoin (Sgórr Deas)	NC 104064	15	E439	
649	Beinn nan Caorach	NC 080053	15	E439	
648	Beinn nan Caorach Beag	NC 087053	15	E439	
705	Sgurr an Fhidhleir	NC 094054	15	E439	
717	Speicein Cóinnich	NC 107042	15	E439	
743	Beinn Mhór na Coigich (Ben Mór Coigach)	NC 094043	15	E439	
738	Garbh Choireachan	NC 087036	15	E439	

Srath-na-Sealga & Fannaich

Many of the mountains of Wester Ross are notable peaks, recognisable from photographs and beloved of climbers. This section covers the area encircled by the A832 to the south and west, and the A835 to the north and east. In addition, the solitary mountain of Beinn Ghobhlach, on the peninsula between Loch Broom and Little Loch Broom is included, a small but perfectly formed summit.

At the northern end of this section is the range of peaks known as An Teallach, comprising of worn Torridonian sandstone. Some of the tops are in excess of 1,000 metres, making them serious mountains, their steep rock sides and narrow ridges only being suitable for experienced walkers. In winter only highly-experienced mountaineers should venture onto these tops. The highest point of An Teallach is Bidein a' Ghlas Thuill, off which are Glas Mheall Mór and Glas Mheall Liath, the latter just missing out on meriting listing as a sperate mountain.

Sgurr Fiona is almost as high as Bidein a' Ghlas Thuill, being just ten feet shorter. It presents itself as a pyramidal mountain, rising between three corries, into which plummet rock cliffs. To the south of Sgurr Fiona a ridge leads to Corrag Bhuidhe, the viewpoint of Lord Berkeley's Seat almost meriting listing as a mountain itself. Beyond rise Stob Cadha Gobhlach and Sáil Liath, two more 3,000 feet peaks.

To the north and west of Sgurr Fiona the mountains diminish in height, here and there rising over 2,000 feet. Amongst those that qualify as mountains in their own right are Sáil Mhór, a prominent summit above Ardessie, which meets Corbett criteria.

The Fannaich group of mountains are located between the A835 and the Corrieshalloch Gorge to the north and Loch Fannich to the south. The latter was a natural loch that was dammed to form a large reservoir for hydro-electric generation. The two highest peaks here are Sgurr Mór and Sgurr nan Clach Geala, which can be climbed together in a route circling Coire Mór. Other Munro tops are included in this round, and the outlying Munro, Meall a' Chrasgaidh can also be added. Further east are some Munro tops, forming considerable climbs in their own right, such as An Coileachan and Beinn Liath Mhór Fannaich, as well as some lower heights. Beinn Liath Mhór a' Ghiubhais Li is a Corbett, rising steeply south of Loch Glascarnoch and climbed from the roadside.

To the west of the Fannaich Forest are A' Chailleach and Sgurr Breac, two high Munros that just fail to reach 1,000 metres. On the far side of Loch a' Bhraoin is Creag Rainich, a Corbett. On the west side of Srath na Sealga from Creag Rainich is Mullach Coire Mhic Fhearchair, rising to 1,019 metres. Around it are a few Munro tops. Sgúrr Bán to the north is a separate Munro, as is Beinn Tarsuinn to the west. These are some of the remotest mountains in Britain, requiring long walks-in to reach.

North of Sgúrr Bán rises Beinn a' Chlaidheimh, a narrow ridge that is virtually 3,000 feet in height. So close to the Munro line is this summit that it was promoted to Munro status in 1974 before being demoted in 2012. Ordnance Survey maps currently indicate the height as 914m, and it is now classed as a Corbett. The Munro Society surveyed the top and found it to be 913.96m in height.

The remote summits of the Fisherfield Forest include A' Mhaighdean and Ruadh Stac Mór, two Munros, as well as Beinn Dearg Mór and Beinn Dearg Bheag, two Corbetts. Beinn a' Chaisgein Mór is another Corbett.

The Letterewe Forest has a series of interesting mountains along the north side of Loch Maree, of which Slioch is the highest and perhaps finest. The northernmost of the twin summit tops is actually the higher. Beinn Lair and Beinn Airigh Chair are two Corbetts on the 'far' side of Loch Maree. To the south east are numerous other lower mountains, as well as the lonely Munro of Fionn Bheinn.

Height	Name	NGR	OS L	OS E	Ascent
635	Beinn Ghobhlach	NH 055943	19	E435	
	Srath-na-Sealga Forest:				
767	△ Sáil Mhór	NH 033887	19	E435	
751	Sgurr Ruadh	NH 040851	19	E435	
1017	Sgurr Creag an Eich	NH 055838	19	E435	
1060	▲ Sgurr Fiona	NH 064837	19	E435	
1049	Corrag Bhuidhe	NH 065833	19	E435	
960	Stob Cadha Gobhlach	NH 069826	19	E435	
954	Sáil Liath	NH 072824	16	E435	
1062	▲ An Teallach (Bidein a' Ghlas Thuill)	NH 069844	19	E435	
979	Glas Mheall Mór	NH 076854	19	E435	
818	Meall Garbh Ard	NH 062865	19	E435	
	Fisherfield Forest:				
820	Beinn Dearg Bheag	NH 020811	19	E435	
906	△ Beinn Dearg Mhór	NH 032799	19	E435	
832 est	Beinn Dearg Mhór Deas	NH 038798	19	E435	

5

5

Height	Name	NGR	OS L	OS E	Ascent
628	Creag-mheall Mór	NG 993815	19	E435	
682	Beinn a' Chaisgein Beag	NG 966822	19	E435	
680	Frith-mheallan	NG 979811	19	E435	
856	△ Beinn a' Chasgain Mór	NG 982785	19	E435	
687	Cadhachan Riabhach	NH 010790	19	E435	
711	Ruadh Stac Beag	NH 027771	19	E435	
705	Meall a' Bhraghad	NH 029763	19	E435	
919	▲ Ruadh Stac Mór	NH 018757	19	E433/E435	
967	▲ A' Mhaighdean	NH 008749	19	E433/E435	
650	Cárnan Ban	NG 999764	19	E435	
652	Beinn Tharsuinn	NG 989744	19	E433	
639	Beinn Tharsuinn Iar	NG 983748	19	E433	
Letterewe Forest:					
653	Meall Chnaimhean	NG 920759	19	E433/E434	
705	Spidean nan Clach	NG 925764	19	E433/E434	
792	△ Beinn Airigh Charr	NG 930762	19	E433/E434	

☐	722	Meall Mheinnidh	NG 955748	19	E433
☐	859	△Beinn Lair	NG 982732	19	E433
☐	808	Sgurr Dubh	NG 988728	19	E433
☐	787	Sgurr Ban	NG 998719	19	E433
☐	684	Meall Daimh	NH 022698	19	E433/E435
☐	934	Sgurr an Tuill Bhain	NH 019689	19	E433/E435
☐	981	▲Slioch (Sleagh)	NH 004691	19	E433/E435
☐	738 est	Sgurr Dubh Mór	NH 010679	19	E433/E435
☐	738	Sgurr Dubh	NH 014675	19	E433/E435
☐	692	Beinn a' Mhuinidh	NH 032660	19	E435
☐	914	△Beinn a' Chlaidheimh	NH 061775	19	E435
☐	989	▲Sgurr Ban	NH 055745	19	E435
☐	654	Meallan an Laoigh	NH 070741	19	E435
☐	1018	▲Mullach Coire Mhic Fhearchair	NH 052735	19	E435
☐	851	Meall Garbh	NH 049726	19	E435
☐	937	▲Beinn Tarsuinn	NH 039728	19	E435
☐	918	Sgurr Dubh	NH 606729	19	E435

5

Height	Name	NGR	OS L	OS E	Ascent
	Loch a' Bhraoin Forest:				
807	△ Creag Rainich	NH 097752	19	E435	
748	Meall Dubh	NH 103747	19	E435	
668	Beinn Bheag	NH 084714	19	E435	
749	Groban	NH 100709	19	E435	
694 est	Ceann Garbh Meallan Chuaich	NH 116697	19	E435	
739	An Sguman	NH 139691	19	E435	
997	▲ A' Chailleach	NH 135714	19	E435	
935	Toman Coinnich	NH 149713	20	E435	
999	▲ Sgurr Breac	NH 158711	20	E435/E436	
	Fannaich Forest:				
934	▲ Meall a' Chrasgaidh	NH 185733	20	E436	
1093	▲ Sgurr nan Clach Geala	NH 184715	20	E436	
922	▲ Sgurr nan Each	NH 185698	20	E436	
868	Sgurr a' Chadha Dheirg	NH 183690	20	E436	
780	Sáil Mhór	NH 177689	20	E436	

The Mountains of Great Britain

	Height	Name	Grid Ref		
☐	961	Cárn na Criche	NH 197726	20	E436
☐	1108	▲Sgurr Mór	NH 203718	20	E436
☐	949	▲Meall Gorm	NH 221696	20	E436
☐	923	▲An Coileachan	NH 242680	20	E436
☐	633	Meallan Buidhe	NH 246695	20	E436
☐	954	▲Beinn Liath Mhór Fannaich	NH 220724	20	E436
☐	664	Beinn Liath Bheag	NH 243737	20	E436

Kinlochluichart Forest:

	Height	Name	Grid Ref		
☐	768	△Beinn Liath Mhór a' Ghiubhais Li	NH 281713	20	E436
☐	686	Beinn Dearg	NH 282684	20	E436
☐	676	Meall na Speireig	NH 301698	20	E436

Loch a' Chroisg Forest:

	Height	Name	Grid Ref		
☐	711	Beinn nan Ramh	NH 140662	19	E435
☐	705	Meall a' Chaorainn	NH 136604	19	E435
☐	933	▲Fionn Bheinn	NH 148621	20	E435

Easter Ross

The most famous mountain of Easter Ross is Ben Wyvis, which rises to the north of Strathpeffer and Dingwall. It comprises Moine schists and gneisses. It is a popular climb, being of Munro status, and located fairly close to some centres of population. The highest point of the mountain is Glas Leathad Mór, at 3,433 feet. The shortest route to the summit is from Garbat to the west, which requires a steep ascent, but various routes from the east, in Glen Glass, are also used. To the north of the main summit are additional Munro tops—Tom a' Choinich and Glas Leathad Beag. Much of the mountainside forms part of the Ben Wyvis National Nature Reserve, noted for its mosses and dotterel breeding populations. To the south-west of Ben Wyvis itself is the smaller Little Wyvis, which qualifies as a Corbett. Ben Wyvis was at one time identified as a possible snow-sports centre, but these proposals have come to nothing. The old *Ordnance Gazetteer of Scotland* noted that 'its upper parts, even in the height of the warmest summers, are almost constantly sheeted or flecked with snow'.

To the north and north-east of Ben Wyvis is a stretch of wild country where a number of summits achieve mountain status. Most of these are quiet tops, though in recent years a number have been disfigured by the construction of wind farms, such as Beinn Tharsuinn and Meall an Tuirc. Further inland some of the summits are a bit higher, including Cárn Chuinneag, a double-topped mountain, the eastmost summit being a Corbett.

The Freevater Forest contains some of the remotest hills in the district, and of these Seana Bhraigh is of Munro status. It has been claimed that this mountain is one of the two most inaccessible Munros in Scotland. The peak of An Sgurr nam Creag an Duine is a distinctive tooth or rock, high above Loch a' Choire Mhóir. Ascents of Seana Bhraigh are usually made from Oykel Bridge to the north-east, which requires a long walk-in through tracks to Strath Mulzie. At the east side of Coire Mór is Cárn Bán, a Corbett in a group of similarly-tall summits, some of which are more interesting, such as Cárn an Toll Lochan (or Creagan a' Chait), and the two Bodachs.

Rising above the east side of Loch Vaich is another Corbett, Beinn a' Chaisteil, with Meall a' Ghrianain only failing to merit Corbett status by a re-ascent of 100 feet.

The Srath a' Bhathaich Forest on the west side of Loch Vaich contains some interesting summits, including the Munro, Am Faochagach. Near to it is Cárn Gorm-loch, which almost reaches 3,000 feet. Loch Toll a' Mhuic was a natural loch that has been extended considerably by the construction of a dam to form what is now Loch Vaich. The waters pass through a tunnel to a power station at Loch Glascarnoch, another reservoir in the Conon scheme of the 1950s.

The highest mountain in this section is Beinn Dearg, which rises to the north of A835 which passes through the Dirrie More from Loch Glascarnoch to the Corrieshalloch Gorge. One of the more common routes of ascent is from Inverlael, at the head of Loch Broom, which takes the walker through the Inverlael Forest and up Gleann na Sguaib to the bealach at the north side of the mountain. A very steep climb brings one to the summit cairn, at 3,547 feet. The view is extensive, with An Teallach to the north-west and Torridon to the west. Coire Gránda on Beinn Dearg's south-east has some impressive schist cliffs, with Creag a' Choire Ghránda being a significant buttress.

Also from this same Bealach an Lochan Uaine, the other Munros of Cona' Mheall, Meall nan Ceapraichean and Eididh nan Clach Geala can be reached. Cona' Mheall is a fine angled summit covered with rocks and boulders. Nearer to the Corrieshalloch Gorge than Beinn Dearg is Beinn Enaiglair, a Corbett. Beinn Dearg is protected as a Special Protection Area for its dotterels and golden eagles. In general, most of these mountains are rather remote.

Height	Name	NGR	OS L	OS E	Ascent
689	Sithean a' Choin Bhain	NH 600803	21	E438	
692	Beinn Tharsuinn	NH 607792	21	E438	
637	Torr Leathann	NH 613785	21	E438	
659	Cnoc an t-Sithein Mór	NH 595782	21	E438	
647	Cárn Salachaidh	NH 518874	20	E437	
646	Cárn an Lochain	NH 502840	20	E437	
739	Cárn Maire	NH 488438	20	E437	
666	Creachan an Fhiodha	NH 494829	20	E437	
839	△ Cárn Chuinneag	NH 484834	20	E437	
829	Cárn Chuinneag Iar	NH 475833	20	E437	
631	Mullach Coire na Gaoitheag	NH 461817	20	E437	
647	Mullach Coire Preas nan Seana-char	NH 440809	20	E437	
668	Creag Ruadh	NH 437817	20	E437	
686	Leaba Bhruic	NH 432836	20	E437	
710	Beinn Tharsuinn	NH 413829	20	E437	
689	Dunan Liath	NH 411849	20	E437	

6

The Mountains of Great Britain

692	Cárn Feur-lochain	NH 405845	20 E437
630 est	Cárn Lochan Sgeireich	NH 391843	20 E437
648	Cárn Crom-loch	NH 388826	20 E437
788	△ Beinn a' Chaisteil	NH 370801	20 E437
772	Meall a' Ghrianain	NH 364776	20 E437
628	Cárn an Aighean	NH 388784	20 E437
697	Cárn Loch nan Amhaichean	NH 412757	20 E437
672	Cárn nan Con Ruadha	NH 414743	20 E437
637	Cárn Choire Bheachain	NH 424735	20 E437
639	Cárn Mór	NH 422716	20 E437

Kildermorie Forest:

743	Beinn nan Eun	NH 448759	20 E437
644	Sgórr a' Chaorainn	NH 464781	20 E437
648	Meall Beag	NH 506752	20 E437
738	Meall Mór	NH 515745	20 E437
626	Meall an Tuirc	NH 540728	20 E437

6

Height	Name	NGR	OS L	OS E	Ascent
	Wyvis Forest:				
643	Queen's Cairn (Cárn an Ban-righ)	NH 466720	20	E437	
704	Meall na Drochaide	NH 509699	20	E437	
928	Glas Leathad Beag	NH 492706	20	E437	
953	Tom a' Choinnich	NH 464700	20	E437	
1046	▲ Glas Leathad Mór (Ben Wyvis)	NH 463683	20	E437	
620	Meall na Speireig	NH 496661	20	E437	
705	Tom na Caillich	NH 439653	20	E437	
763	△ Little Wyvis	NH 430645	20	E437	
	Freevater Forest:				
635	Cárn Alladale	NH 409898	20	E437	
667	Leaba Bhaltair	NH 398914	20	E437/E440	
701	Cárn a' Choin Deirg	NH 398923	20	E437/E440	
745	An Socach	NH 378868	20	E437	
833	Bodach Mór	NH 361887	20	E437	
837	Bodach Beag	NH 355878	20	E437	

6

842	☐	△Cárn Ban	NH 339875	20	E437
778	☐	Cárn an Toll Lochan	NH 325880	20	E437
736	☐	Cárn Loch Srúban Móra	NH 320848	20	E436
794	☐	Stob Coire Mór	NH 306871	20	E436
905	☐	An Sgurr nam Creag an Duine	NH 297879	20	E436
905	☐	Seana Bhraigh Beag	NH 287872	20	E436
926	☐	▲Seana Bhraigh	NH 281879	20	E436
679	☐	Meall nam Bradhan	NH 267903	20	E436
		Srath a' Bhathaich Forest:			
632	☐	Meall a' Chaorainn	NH 360827	20	E437
629	☐	Meall a' Chuaille	NH 343821	20	E437
910	☐	Cárn Gorm-loch	NH 319801	20	E436
953	☐	▲Am Faochagach	NH 304794	20	E436
761	☐	Srón Liath	NH 306763	20	E436
742	☐	Tom Ban Mór	NH 318753	20	E436
		Inverlael Forest:			
646	☐	Meall Dubh	NH 214900	20	E436

6

Height	Name	NGR	OS L	OS E	Ascent
667	Beinn Bhreac	NH 226887	20	E436	
648	Cárn Mór	NH 247870	20	E436	
858	Meall Glac an Ruidhe	NH 265862	20	E436	
804	Meall a' Choire Ghlais	NH 273859	20	E436	
862	Toman Coinich	NH 272844	20	E436	
873	An Socach	NH 257852	20	E436	
927	▲ Eididh nan Clach Geala	NH 258842	20	E436	
883	Cnap Coire Loch Tuath	NH 282828	20	E436	
977	▲ Meall nan Ceapraichean	NH 257826	20	E436	
978	▲ Cona' Mheall	NH 274816	20	E436	
655	Cárn Loch nan Eilean	NH 273788	20	E436	
995	Creag a' Choire Ghranda	NH 267799	20	E436	
1084	▲ Beinn Dearg	NH 259812	20	E436	
874	Iorguill	NH 239816	20	E436	
809	Creag Bac na Faire	NH 228807	20	E436	
890	△ Beinn Enaiglair	NH 225805	20	E436	

6

6

Meall Doire Faid	NH 221792	20	E436
Meall nan Doireachan	NH 229783	20	E436
Meall Leachachain	NH 245771	20	E436

729 713 621

Torridon

Some of the mountains of Torridon are the finest in Britain, and since they rise from almost sea level in many cases make considerable climbs. Most of this section, which includes the group of summits within the Torridon Forest, north of the A896, as well as the lesser-known peaks south of that road and also the summits of Applecross, fall within the Wester Ross National Scenic Area.

It is the northern mountains that are the most impressive to most climbers. North of Glen Torridon are two icons of the Scottish Highlands—Liathach and Beinn Eighe. The former, and the south-west slopes of Beinn Eighe are owned by the National Trust for Scotland, one of the mountainous properties acquired by the trust due to the generosity of Percy Unna. Liathach is a ridge of summits composed of Torridonian sandstone, of which Spidean a' Choire Léith is the tallest. There are six mountain peaks in the range, of which two are Munros—the tallest summit and also Mullach an Rathain. Am Fasarinen Mór is the highest of The Pinnacles, a tricky section of the mountain.

East of Liathach is Beinn Eighe, located within the Beinn Eighe National Nature Reserve. It, too has ancient sandstone in its make up, but the summit has a Cambrian quartzite cap. The ridge of five mountains makes a great climb, but the highest point of the mountain is Ruadh-stac Mór, located on a ridge perpendicular to the main spine. Like Liathach, this mountain has two Munros in its make up.

The third great Torridonian mountain is Beinn Alligin, which rises west of Liathach and north of Loch Torridon. It also falls within the National Trust for Scotland's Torridon estate. Two Munros make up this mountain, plus a third top, Sgurr Beag, sometime known as Na Rathanan or An t-Sáil Bheag. The views from the summit on a clear day are regarded as some of Scotland's finest.

North and west of the high Torridon summits are the Shieldaig and Flowerdale forests, wild open countryside with a few spectacular peaks, though failing to reach Munro status they are less often climbed. Baosbheinn is a group of four summits, whereas Beinn an Eóin is a ridge with two main tops. Beinn a' Chearcaill and Meall a' Ghiuthais are two more notable peaks. In the midst of this area rises Beinn Dearg, a significant mountain that is around 3,000 feet in height. Historical maps indicate that it was 2,995 feet above sea level, whereas metric maps show it as 914 metres tall. It was classed as a Munro for a short time, but new surveys have returned it to sub-Munro height, and therefore it has re-joined the list of Corbetts.

To the south of Glen Torridon are a series of fine mountains, of which only three, Maol Chean-dearg, Sgórr Ruadh and Beinn Liath Mhór are Munros. Although reachable from Torridon, most ascents of these peaks are made from Achnashellach or elsewhere in Glen Carron to the south. Failing to reach Munro status by around thirty feet or so, is Beinn Damh, an attractive peak in its own right, and one of five Corbetts in the area. The others are An Ruadh Stac, Fuar Tholl, Sgórr nan Lochan Uaine and Sgurr Dubh.

West of the A896 is Applecross, where a high ridge separates the remote communities overlooking the Inner Sound from the mainland. The famous Bealach na Ba hill road crosses these mountains, allowing the southern summits to be reached fairly easily from a public road. Meall Gorm is to the south of the pass head, with extensive views to the Cullin on Skye. Sgurr a' Chaorachain rises on the north side, where a track to a transmitter makes climbing its north top fairly simple. The main summit (a Corbett) requires walking along a narrow, but simple ridge, from where views over to Kintail are possible.

The highest Applecross summit is Beinn Bhán, a Corbett, which has a series of steep corries on its eastern flanks. Ascents are often made from Tornapress or Couldoran area.

7

Height	Name	NGR	OS L	OS E	Ascent
	Shieldaig Forest:				
624	Beinn Bhreac	NG 837640	19/24	E433	
675	Creag a' Chinn Duibh	NG 859629	19/24	E433	
705	Baosbheinn - Ceann Beag	NG 882644	19/24	E433	
806 est	Baosbheinn Meadhonach	NG 877651	19/24	E433	
875	△ Baosbheinn - Sgòrr Dubh	NG 871653	19/24	E433	
841 est	Baosbheinn - Sgòrr Dubh North Top	NG 867657	19/24	E433	
801	Baosbheinn - Stob Coire Beag	NG 862669	19/24	E433	
	Flowerdale Forest:				
715	Beinn Tuath an Eoin	NG 897660	19/24	E433	
855	△ Beinn an Eoin	NG 905646	19	E433	
725	Beinn a' Chearcaill	NG 931638	19	E433	
641	Creag na Feol	NG 943641	19	E433	
887	△ Meall a' Ghiubhais	NG 978638	19	E433	
	Torridon Forest:				
672	An Ruadh-mheallan	NG 836615	19/24	E433	

The Mountains of Great Britain

	Height	Name	Grid Ref	Map
☐	922	▲ Beinn Alligin - Tom na Gruagaich	NG 859602	19/24 E433
☐	986	▲ Beinn Alligin - Sgurr Mhór	NG 866613	19/24 E433
☐	866	Beinn Alligin - Sgurr Bheag	NG 873613	19/24 E433
☐	889	Stúc Loch na Cabhaig	NG 891616	19/24 E433
☐	914	△ Beinn Dearg	NG 894608	19/24 E433
☐	761	Cárn na Feola	NG 915613	19/24 E433
☐	955	Meall Dearg	NG 914580	25 E433
☐	1023	▲ Mullach an Rathain	NG 912677	25 E433
☐	909 est	Am Fasarinen Mór	NG 923574	25 E433
☐	1055	▲ Liathach - Spidean a' Choire Léith	NG 929580	25 E433
☐	983	Stob a' Choire Liath Mhór	NG 933582	25 E433
☐	915	Stúc a' Choire Dhuibh Bhig	NG 942582	25 E433
☐	980	Sáil Mhór	NG 938605	19 E433
☐	976	A' Choinneach Mór	NG 945600	19 E433
☐	1010	▲ Ruadh Stac Mór	NG 952612	19 E433
☐	993	▲ Spidean Coire nan Clach	NG 965596	25 E433
☐	896	△ Ruadh Stac Beag	NG 973613	19 E433

7

Height	Name	NGR	OS L	OS E	Ascent
970	Sgurr Ban	NG 974600	19	E433	
963	Sgurr an Fhir Duibhe	NG 982600	19	E433	
	Leadgobhainn Forest:				
625	Beinn na Feusaige	NH 090543	25	E429	
678	Càrn Breac	NH 045530	25	E429	
	Cuilinn Forest:				
782	△Sgurr Dubh	NG 979557	25	E429/E433	
661 est	Cruach Loch nan Corrag	NG 975553	25	E429/E433	
871	△Sgórr nan Lochan Uaine	NG 969532	25	E429	
737	Beinn Liath Bheag	NG 985527	25	E429	
926	▲Beinn Liath Mhór	NG 964520	25	E429	
887	Beinn Liath Mhór Meadhonach	NG 975518	25	E429	
876	Beinn Liath Mhór Ear	NG 983515	25	E429	
769	Beinn Mheadhoin	NG 957514	25	E429	
962	▲Sgórr Ruadh	NG 959505	25	E429	
904	Sgórr Raeburn	NG 964507	25	E429	

7

895	Stob a' Choire Mainreachan	NG 972488	25	E429
907	△Fuar Tholl	NG 975489	25	E429
646	Meall Dearg	NG 933518	25	E429
933	▲Maol Chean-dearg (Meall a' Chinn Deirg)	NG 924498	25	E429
677	Meall nan Ceapairean	NG 937486	25	E429
892	△An Ruadh-stac	NG 922481	25	E429
729	Glas Bheinn	NG 902441	25	E428/E429
706	Creag na h-Iolaire	NG 899444	24	E428/E429
732	Sgurr a' Gharaidh	NG 884443	24	E428/E429

Beinn-damh Forest:

736	Beinn na h-Eaglaise	NG 909524	25	E429
676	Leac Dhubh	NG 906518	25	E429
687	Sgurr nan Bana Mhóraire	NG 870527	24	E428/E429
675	Meall Gorm	NG 874523	24	E428/E429
868	Beinn Damh Beag	NG 887510	24	E428/E429
903	△Beinn Damh	NG 893502	24	E428/E429

7

7

Height	Name	NGR	OS L	OS E	Ascent
	Applecross Forest:				
710	Meall Gorm	NG 779410	24	E428	
773	Sgurr Iar a' Chaorachain	NG 786424	24	E428	
792	△Sgurr a' Chaorachain	NG 797417	24	E428	
651	Creag an Eas	NG 786432	24	E428	
673	Cárn Dearg	NG 783450	24	E428	
763	Stob Coire Each	NG 814439	24	E428	
752 est	A' Chioch	NG 812447	24	E428	
896	△Beinn Bhan	NG 803450	24	E428	
674	A' Phoit	NG 809453	24	E428	
846	Srón Coire an Fhamhair	NG 804464	24	E428	
626	Meall an Doireachain (Beinn a' Chlachain)	NG 723491	24	E428	

54

Monar & Strath Conon

Forming a large part of the North West Highlands, this section is bounded by the A832/A890 to the north and the Glen Cannich/Loch Mullardoch/Glen Elchaig gap to the south. The area is divided by deep glens which make their way into the hills from the east—Strathconon, Glen Orrin, and Glen Strathfarrar, which make good routes for access, though the Glen Strathfarrar road is private.

At the north-eastern end, the mountains of Strathconon Forest are substantial summits, with steep sides and some degree of prominence. Of these, Bac an Eich reaches Corbett proportions, a substantial mountain south of Scardroy and Loch Beannacharain. Other considerable peaks lie to the north of Strath Conon itself, including Meallan nan Uan and Sgurr a' Mhuilinn, a pair of Corbetts often ascended together in a walk around Allt an t-Srathain Mhóir.

On the south side of Glen Orrin, and to the north of Glen Strathfarrar, is a range of mountains that includes four Munros. A couple of these are in excess of 1,000 metres—Sgurr Fhuar-thuill and Sgurr a' Choire Ghlais. A couple of other summits rise in excess of 1,000 metres. Two other Munros lie to the east—Cárn nan Gobhar and Sgurr na Ruaidhe. Beyond this the land drops to a lower level, but with a few summits in the marshy expanse, before rising to the Corbett, Benn a' Bha'ach Ard. This is often climbed on a circular route from Inchmore over Sgurr a' Phollain and other tops.

The East Monar Forest is one of the more remote parts of the Highlands, requiring a long walk-in to reach the summits. The highest mountain hereabouts is Maoile Lunndaidh, which is just over 1,000 metres in height. Its bald summit is enlivened by the deep corries around it, especially Fuar-tholl Mór. West of a deep glen rises Bidean an Eóin Deirg and Sgurr a' Chaorachain, another pair of 1,000 metre tops. These are often climbed from Glen Carron to the

8

north, the track up the side of the Allt a' Chonais making a useful means of access. Climbing the hills from the south requires a much longer walk, but one that is perhaps more rewarding for its remoteness. Another path from near to Loan in Glen Carron can be used to reach the Munro named Moruisg, which rises south of Loch Sgamhain. Other summits are located in the same range, including Sgurr nan Ceannaichean at the western end. This summit almost reaches 3,000 feet in height, but accurate surveys have proved that it fails to make Munro status by three feet. It was a Munro for a short period between 1981 and 2009 when maps indicated its height as 915 metres, obtained from aerial photogrammetry. More recent surveys confirmed its status as a Corbett.

The East Benula Forest area is another remote stretch of mountain landscape, preserved as a deer forest. Sgurr na Lapaich is the highest mountain in the immediate area, reaching 1,150 metres. Most ascents of this mountain are made from the dam at the east end of Loch Mullardoch, often including Cárn nan Gobhar on the way there or back. An Riabhachan and An Socach are two Munros of considerable height to the west of here, requiring lengthy walks in. Another pair of remote Munros are Lurg Mhór and Bidein a' Choire Sheasgaich. Beinn Dronaig is a substantial Corbett south-west of these summits, all of which are often ascended from Bendronaig Lodge. South of here is the Killilan Forest, where three Corbetts of some significance can be found.

A number of the lochs within this section were dammed for hydro-electric purposes in the 1960s, the waters often tunnelled to power stations located in the glens. Loch Mullardoch at the south incorporates the natural lochs of Mullardoch and Lungard (that part north of Beinn Fhionnlaidh). Loch Monar was also extended, the waters submerging two former shooting lodges. Loch Orrin was created by two dams. Most of the area within this section forms private deer forests and sporting estates. Lower Glen Strathfarrar forms a national nature reserve.

Height	Name	NGR	OS L	OS E	Ascent
	Cabaan Forest:				
☐ 673	Meall nan Damh	NH 352522	26	E431	
☐ 653	Meall a' Bhogair Beag	NH 338524	26	E431	
☐ 672	Meall a' Bhogair Mór	NH 330521	26	E431	
☐ 651	Cárn Uilleim	NH 333515	26	E431	
	Strathconon Forest:				
☐ 673	Cárn na Coinnich	NH 324511	26	E431	
☐ 671	Meall Doire Fheara	NH 314524	26	E430	
☐ 663	Meall Giubhais	NH 306507	26	E430	
☐ 665	Beinn Mheadhoin	NH 259478	25	E430	
☐ 849	△ Bac an Eich	NH 222489	25	E430	
☐ 770	Creag Coire na Feola	NH 206495	25	E430	
☐ 701 est	Meall Buidhe	NH 228503	25	E430	
☐ 680 est	Meall na Faochaig	NH 258525	25	E430	
☐ 734	Creag Ruadh	NH 277539	25	E430	
☐ 838	△ Meallan nan Uan	NH 263544	25	E430	

8

The Mountains of Great Britain

Height	Name	NGR	OS L	OS E	Ascent
879	△ Sgurr a' Mhuilinn	NH 264557	25	E430	
844	Sgurr a' Ghlas Leathaid	NH 244564	25	E430	
848	Sgurr a' Choire-rainich	NH 247569	25	E430	
	East Monar Forest:				
743	Meall Innis na Sine	NH 188482	25	E430	
789	Sgurr Coire nan Eun	NH 198469	25	E430	
814	△ An Sithean	NH 171454	25	E430	
689	Meall Dubh na Caoidhe	NH 183435	25	E430	
613	Druim Dubh	NH 203447	25	E430	
627	Creag Dhubh Bheag	NH 157472	25	E430	
853	Creag Dhubh Mhór	NH 140473	25	E430	
1005	▲ Maoile Lunndaidh	NH 135459	25	E430	
	Gleanfhiodhaig Forest:				
700	Creag Sgiathan	NH 169509	25	E430	
875	Cárn Gorm	NH 135500	25	E430	
828	Stob Toll a' Ghobhain	NH 125500	25	E430	

8

The Mountains of Great Britain

		Height	Name	Grid Ref	Map
☐	851		Stob Coireag nam Mang	NH 121496	25	E430
☐	872		Stob Coire Beithe	NH 112496	25	E429/E430
☐	926	▲	Moruisg	NH 102500	25	E429/E430
☐	913	△	Sgurr nan Ceannaichean	NH 087481	25	E429

West Monar Forest:

☐	1045		Bidean an Eoin Deirg	NH 103443	25	E429
☐	1053	▲	Sgurr a' Chaorachain	NH 088447	25	E429
☐	999	▲	Sgurr Choinnich	NH 077446	25	E429
☐	901 est		Sgurr na Conbhaire	NH 081433	25	E429
☐	863	△	Beinn Tharsuinn	NH 056433	25	E429
☐	795		Beinn Tharsuin Iar	NH 043430	25	E429
☐	945	▲	Bidein a' Choire Sheasgaich	NH 049412	25	E429
☐	987	▲	Lurg Mhór	NH 065404	25	E429
☐	974		Meall Mór	NH 072406	25	E429

Achadh-na-seileach Forest:

☐	819		Sgurr Tuath na Feartaig	NH 054464	25	E429
☐	862	△	Sgurr na Feartaig	NH 054454	25	E429

8

Height	Name		NGR	OS L	OS E	Ascent
805	Sgurr na Feartaig Iar		NH 038449	25	E429	
690	Eagan		NH 020451	25	E429	
612	Creag Dhubh Mhór		NG 983404	25	E429	
882	Càrn Eitege		NH 210432	25	E430	
890	Sgurr na Muice		NH 226417	25	E430	
695	Beinn na Muice		NH 219403	25	E430	
1049	▲ Sgurr Fhuar-thuill		NH 236437	25	E430	
1030	Creag Ghorm a' Bhealaich		NH 244434	25	E430	
1083	▲ Sgurr a' Choire Ghlais		NH 259430	25	E430	
755	Meall an Gheur-fheadain		NH 264413	25	E430	
992	▲ Càrn nan Gobhar		NH 273439	25	E430	
993	▲ Sgurr na Ruaidhe		NH 289427	25	E430	
854	Garbh-charn		NH 292411	25	E430	
913	Aonach na Reise		NH 303429	26	E430	
756	An Leth-chreag		NH 300452	25/26	E430	
686	Sgurr na Cairbe		NH 306463	26	E430	

8

☐ 737	Càrn Ban	NH 336418	26	E431
☐ 766	Meallan Buidhe	NH 337447	26	E431
☐ 862	△Beinn a' Bha'ach Ard (Beinn a' Bhathaich Ard)	NH 360434	26	E431

Coire-thollaidh Forest:

☐ 855	Sgurr a' Phollain	NH 365445	26	E431
☐ 736	Càrn a' Ghorm-lochan	NH 370459	26	E431
☐ 729	Sgurr a' Ghlaisein	NH 360460	26	E431

Baile-mór Forest:

☐ 677	Càrn Gorm	NH 329355	26	E431
☐ 655	Sgórr na Ruadhraich	NH 317359	26	E430
☐ 638	Sgór an Uillt Ghiubhais	NH 317350	26	E430

Glencannich Forest:

☐ 684	Meall a' Mhadaidh	NH 305370	26	E430
☐ 818	△Sgórr na Diollaid	NH 282363	25	E430
☐ 777	Stob Coire na Feithe Seilich	NH 273359	25	E430
☐ 680	An Soutar	NH 259350	25	E430
☐ 694	Meallan Odhar	NH 250357	25	E430

8

Height	Name	NGR	OS L	OS E	Ascent
743	Mullach Tarsuinn	NH 235339	25	E430	
861	Creag Feusaig	NH 218340	25	E430	
946	Creag Dubh	NH 200351	25	E430	
993	▲ Cárn nan Gobhar	NH 182343	25	E430	
East Benula Forest:					
1150	▲ Sgurr na Lapaich	NH 161351	25	E430	
1093	Sgurr nan Clachan Geala	NH 161342	25	E430	
792	Mullach a' Ghlas-thuill	NH 163319	25	E415/E430	
1129	▲ An Riabhachan	NH 133344	25	E430	
1040	Stob Coire Riabhachain	NH 117338	25	E429/E430	
1069	▲ An Socach	NH 101353	25	E429/E430	
706	An Cruachan	NH 094359	25	E429	
619	Beinn Bheag	NH 106376	25	E429/E430	
679	Cárn na Breabaig	NH 067302	25	E414/E429	
797	△ Beinn Dronaig	NH 037382	25	E429	
Killilan Forest:					

8

	Height	Name	Grid Ref		Map
☐	870	An Creachal Beag	NH 067332	25	E429
☐	899	△Aonaich Buidhe	NH 058325	25	E414/E429
☐	868	△Faochaig	NH 022317	25	E414/E429
☐	724	Srón na Gaoithe	NH 008306	25	E414/E429
☐	656	Creag nan Eilid	NH012296	25	E414
☐	879	△Sguman Coinntich	NG 977304	25	E414/E429
☐	788	Cárn a' Bhealach Mhic Bheathain	NG 984311	25	E414/E429
☐	755 est	Srón a' Choire Dail Aiteil	NG 984317	25	E414/E429
☐	754	Ben Killilan	NG 975317	25	E414/E429
☐	753	Sgurr na Cloiche (Beinn Killilan)	NG 961316	25	E414/E429

8

Glen Affric & North Kintail

The peaks of Glen Affric and North Kintail are some of the better-known in Scotland. Who has never seen the view (at least on paper) from Mám Rattachan or thereabouts across Loch Duich to the Five Sisters of Kintail, more properly known as Beinn Mhór? Or what of the view of Sgurr na Lapaich across Loch Affric? Failing that, one must have seen a view of Eilean Donan Castle, which stands on its islet at the western end of this district. The scenery, as one can imagine, is some of the finest in the highlands, Glen Affric regarded by many as the most beautiful in Scotland.

Glen Affric (or Affaric, as the estate prefers it) begins near Cannich and stretches westwards into the hills. Loch Beinn a' Mheadhoin is the first loch reached along this road, attractive in its forest setting, enlarged by the construction of a hydro-electric dam, but on reaching Loch Affric the true splendour of the western highlands becomes apparent. Beyond the head of the loch the glen is less frequented, only walkers heading for the hostel at Alltbeithe to be seen. Pathways continue by way of two glens to either Loch Duich or else Loch Long by way of the Falls of Glomach.

From Loch Affric an ascent can be made of Mám Sodhail and Cárn Eighe; many different route possibilities presenting themselves. Cárn Eighe is the highest peak north-west of the Great Glen, and thus, due to its re-ascent, can claim to be Scotland's second most significant mountain. Sgurr nan Ceathreamhnan has twin peaks, the eastern one the highest, and it can be reached by a path from Alltbeithe into Coire na Cloiche and westward along the ridge. Alternatively a way can be made to the top from the Falls of Glomach or else from Loch Duich through Bealach an Sgairne to Loch a' Bhealacih, at the foot of Sgurr Gaorsaic.

Beinn Fhada and the Five Sisters are owned by the National Trust for Scotland, part of the mountain country bought from the Percy Unna fund in 1944 to add to land gifted in 1941 containing the Falls of

Glomach, which tumble 370 feet from a hanging valley. The trust lands include the site of the Battle of Glenshiel, fought on 10 June 1719 between government troops and about 300 Spaniards who landed at Loch Duich to fight for the Jacobite cause. The Spanish connection gives rise to Sugrr na Spainteach, *peak of the Spaniards*, immediately north of the battlefield.

Beinn Fhada is best climbed from Morvich, where there is parking, a popular camp-site and outdoor centre. A track ascends Gleann Lichd to Glenlicht House and two bridges crossing the River Croe and Allt Grannda. The summit is a straight, but steep, climb from here. The path continues past the falls on the Allt Grannda to Alltbeithe hostel, passing Camban bothy en route.

The Five Sisters are best climbed in a single journey, with transport to the Bridge of Shiel, but it is quite an undertaking, with over ten thousand feet of ascent. A similar one-way journey is preferable to climb the three Munros east of the Five Sisters—Sgurr an Fhuarail, Sgurr a' Bhealaich Dheirg and Sáileag. Some walkers like to include a fourth summit, Ciste Dhubh, from where a descent can be made to the path through An Caoran Mór to Loch Cluanie. East of this pass is A' Chralaig and Mullach Fraoch-choire, two more Munros, the former being climbed by Bonnie Prince Charlie on his travels after Culloden.

There is another group of high summits in the Ceannacroc Forest, north of Loch Cluanie, of which Sgurr nan Conbhairean is the tallest. A path from Lundie, at first following a military road, winds its way up the shoulder of Cárn Ghluasaid, a Munro, and then along the ridge to Conbhairean. In Coire Mheadhon, east of Tir Mór na Seilge, also known as Sáil Chaorainn, is Prince Charlie's Cave, one of his many hideouts.

The summits in the Balmacaan Forest are rather low, though in some places rocky, Meall Fuar-mhonaidh above Loch Ness being the only one which is well-known.

Height	Name	NGR	OS L	OS E	Ascent
	Fasnakyle Forest:				
646	Beinn a' Chairean	NH 296317	25	E415/E430	
638	Beinn na Cuidhe	NH 285309	25	E415/E430	
663	Creag a' Choire Duibh	NH 273312	25	E415/E430	
716	Càrn Loch na Gobhlaig	NH 257302	25	E415/E430	
693	Meall Mór	NH 249281	25	E415/E430	
678	Feith a' Ghiubhais	NH 240297	25	E415	
892	Doire Tana	NH 219282	25	E415	
1054	▲Toll Creagach	NH 194283	25	E415	
745	Creag a' Bhaca	NH 207296	25	E415	
613	Beinn a' Mheadhoin	NH 219256	25	E415	
653	Am Meallan	NH 191248	25	E415	
1112	▲Tom a' Choinich	NH 163273	25	E415	
1051	An Leth-chreag	NH 153270	25	E414/E415	
1131	Srón Gharbh	NH 144263	25	E414	
1147	Stob a' Choire Dhomhain	NH 132264	25	E414	

9

The Mountains of Great Britain

	Height	Name	Grid ref	Map	Explorer
☐	913	Creag na h-Eige Beag	NH 141283	25	E414
☐	1183	▲ Cárn Eighe	NH 123262	25	E414
☐	1005	▲ Beinn Fhionnlaidh	NH 116283	25	E414
☐	1181	▲ Mám Sodhail	NH 120253	25	E414
☐	996	Mullach Cadha Rainich	NH 139247	25	E414
☐	1036	Sgurr na Lapaich	NH 154244	25	E414/E415
☐	1074	An Tudair	NH 127240	25	E414
☐	921	▲ An Socach	NH 088230	25/33	E414
☐	915	Stob Coire na Cloiche	NH 075227	25/33	E414
☐	1151	▲ Sgurr nan Ceathreamhnan	NH 057228	25/33	E414
☐	967	Cárn na Con Dubh	NH 072242	25/33	E414
☐	982	▲ Mullach na Dheiragain	NH 081259	25/33	E414
☐	974	Mullach Sithidh	NH 082264	25/33	E414
☐	1143	Sgurr Iar nan Ceathreamhnan	NH 053228	25/33	E414
☐	1075	Stúc Bheag	NH 053238	25/33	E414
☐	1041	Stúc Mór	NH 053243	25/33	E414
☐	856	Creag Ghlas	NH 047264	25/33	E414

9

9

Height	Name	NGR	OS L	OS E	Ascent
657	Sgurr na h-Eige	NH 052276	25/33	E414	
839	△Sgurr Gaorsaic	NH 037219	25/33	E414	
631	Boc Mór	NG 918259	25/33	E413	
634	Cárn Bad a' Chreamha	NG 926265	25/33	E413	
841	Sgurr an Airgid	NG 941228	25/33	E413/E414	
703	Beinn Bhuidhe	NG 961235	25/33	E413/E414	
675	Beinn Bhreac	NG 973248	25/33	E414	
729	Cárnan Cruithneachd	NG 994258	25/33	E414	
918	▲A' Ghlas-bheinn	NH 008231	25/33	E414	
782	Meall a' Bhealaich	NH 012211	25/33	E414	
870	Faradh Nighean Fhearchair	NG 995209	25/33	E414	
954	Meall an Fhuarain Mhóir	NH 000196	33	E414	
1032	▲Beinn Fhada (Ben Attow)	NH 018193	33	E414	
962	Sgurr a' Duibhe Doire (Sgurr a' Dubh Doire)	NH 034185	33	E414	
876	Sgurr na Moraich	NG 965193	33	E413/E414	
869	Beinn Bhuidhe	NG 969182	33	E413/E414	

The Mountains of Great Britain

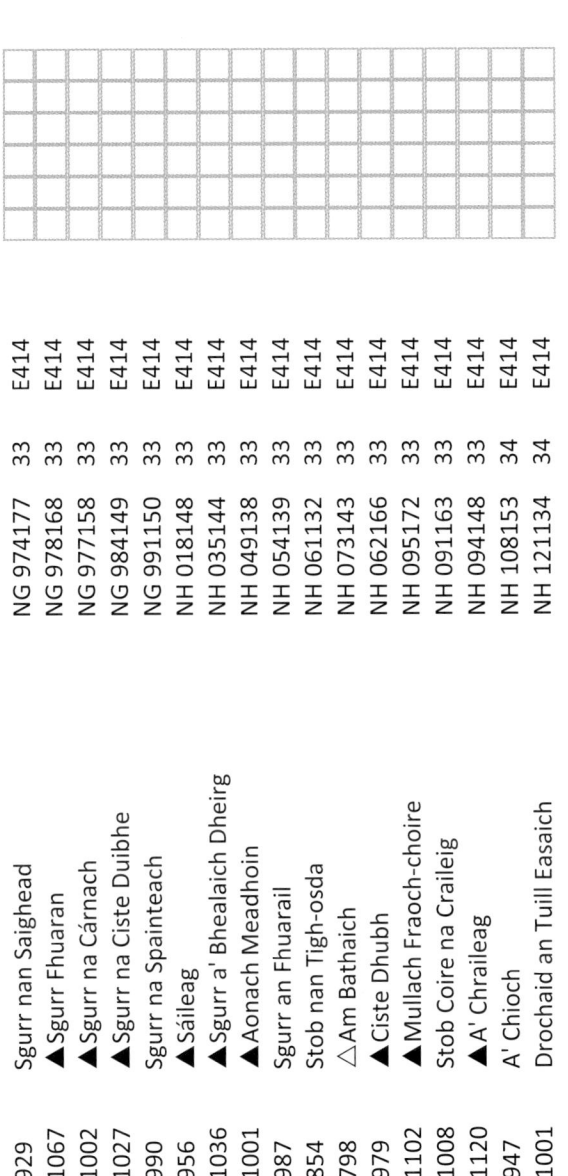

929	Sgurr nan Saighead	NG 974177	33	E414
1067	▲ Sgurr Fhuaran	NG 978168	33	E414
1002	▲ Sgurr na Cárnach	NG 977158	33	E414
1027	▲ Sgurr na Ciste Duibhe	NG 984149	33	E414
990	Sgurr na Spainteach	NG 991150	33	E414
956	▲ Sáileag	NH 018148	33	E414
1036	▲ Sgurr a' Bhealaich Dheirg	NH 035144	33	E414
1001	▲ Aonach Meadhoin	NH 049138	33	E414
987	Sgurr an Fhuarail	NH 054139	33	E414
854	Stob nan Tigh-osda	NH 061132	33	E414
798	△ Am Bathaich	NH 073143	33	E414
979	▲ Ciste Dhubh	NH 062166	33	E414
1102	▲ Mullach Fraoch-choire	NH 095172	33	E414
1008	Stob Coire na Craileig	NH 091163	33	E414
1120	▲ A' Chraileag	NH 094148	33	E414
947	A' Chioch	NH 108153	34	E414
1001	Drochaid an Tuill Easaich	NH 121134	34	E414

9

69

Height	Name	NGR	OS L	OS E	Ascent
1109	▲ Sgurr nan Conbhairean	NH 130139	34	E414	
998	Creag a' Chaorainn	NH 137132	34	E414	
957	▲ Cárn Ghluasaid	NH 146125	34	E414	
732	Cárn nam Feuaich	NH 175125	34	E415	
845	An Reithe	NH 153150	34	E414/E415	
1002	▲ Sáil Chaorainn	NH 133154	34	E414	
1001	Cárn na Coire Mheadhoin	NH 134159	34	E414	
929	Tigh Mór na Seilge	NH 140167	34	E414	
773	Beinn an Iomaire	NH 162165	34	E415	
865	△ Cárn a' Choire Ghairbh	NH 137189	34	E414	
771	Cárn Glas Iochdarach	NH 162202	25	E415	
661	Creag nan Calman	NH 200206	25	E415	
724	Cnap na Stri	NH 190197	34	E415	
872	Cárn nan Coireachan Cruaidh	NH 186179	34	E415	
888	△ Aonach Shasuinn	NH 173180	34	E415	
847	Cárn a' Choire Bhuidhe	NH 184170	34	E415	

9

612	Meallan Odhar	NH 211173	34	E415
706	Cárn a' Chaochain	NH 235178	34	E415
648	Cárn a' Choire Leith	NH 264189	34	E415
664	Cárn nan Earb	NH 301195	34	E415
678	Cárn Mhic an Toisich	NH 311185	34	E415
610	Cárn na Caorach	NH 323187	34	E415
632	Meall a' Chrathaich Beag	NH 353216	26	E415
679	Meall a' Chrathaich	NH 361221	26	E415
639	Cárn a' Mhuilinn	NH 365223	26	E415
615	Meall nan Oighreagan	NH 372227	26	E416
613	Cárn na Ruighe Duibhe	NH 376240	26	E416
616	Cárn Tarsuinn	NH 387223	26	E416
651	Glas-bheinn Mhór	NH 437232	26	E416
699	Meall Fuar-mhonaidh	NH 457223	26	E416

9

Monadh Liath

The Monadhliath Mountains form one of the vastest and most desolate stretches of upland in the British Isles, although in recent years wind farms and hydro-electric works have left their mark. The name is Gaelic for 'grey moor', an apt description. Only at the southern end, in the summits of Creag Meagaidh and Carn Bán district, do the hills take on a mountainous aspect, elsewhere being but a great elevated plateau of moorland. Only the dissections of valleys, such as Glen Markie or Glen Tarff, create steep sides to the hills, elsewhere gentle slopes predominate. One could walk for miles across moors and see no-one, so remote is this mountain range. Three of the north's great rivers rise amid these hills—Spey at Loch Spey, Nairn and Findhorn.

The Cárn Bán group of Munros can be climbed in a single horseshoe walk. From Newtonmore a roadway heads up Glen Banchor, turning into a path following the River Calder into Coire nan Laogh and Loch Dubh. Cárn Dearg lies at the head of the corrie, Cárn Bán a short distance to the north of it. The estate boundary can be followed over Cárn Ballach to Cárn Sgulain, which must be one of the most undistinguished Munros ever. A' Chailleach is due south of this. A route can be made over the rough slopes to the Allt a' Chaorainn and back to Newtonmore. Geal Charn can be reached from either Crathie or Garbha Bridge, pathways leading up glens at the Munro's feet. The route from Crathie into Glen Markle, and up to Lochan a' Choire, is of more interest than that from Garbha Bridge.

Creag Meagaidh, rising above Loch Laggan, is the most spectacular peak in the district, as well as being the tallest. Its lower slopes, Coire Ardair and the Moy Forest, comprise of birch woods which are now protected as a nature reserve. Hidden from public eyes are the great corries of the mountain, Coire Ardair being the greatest, with its mile-long rock headwall. A number of Munro tops make up Creag Meagaidh's

10

probing fingers. This mountain, Moy and Aberarder forests were acquired by the Nature Conservancy Council to prevent the commercial afforestation of the lower slopes. The reserve extends to about 10,000 acres.

Beinn a' Chaorainn and Beinn Teallach are both Munro summits, the latter promoted from Corbett status. A track from Roughburn climbs part of the lower slopes of Beinn a' Chaorainn, and gives access to a path which follows the Allt a' Chaorainn, from which Beinn Teallach is a straightforward climb up its southern shoulder.

Westwards is Glen Roy, also a National Nature Reserve, famed for its parallel roads, the creation of great lochs formed when Glen Spean was blocked with ice. As the ice slowly melted, the level of these lochs subsided gradually, each level creating a beach of small stones which now make up these 'roads'. Neighbouring Gleann Glaoidh also has this unique phenomenon. The two Cárn Deargs at the head of the glen are both Corbetts, as is Beinn Iaruinn.

Crossing the middle of the Monadh Liath is the Corrieyairack Pass, traversed by a military road built by General Wade in 1731. The track climbs to a height of 2,507 feet at its highest, and makes a popular walk. The track also gives good access to Cárn Leac, Cnoc a' Coire Dherrig and Gairbeinn, the latter a Corbett.

Away to the north-east of this the mountains are at their most remote, only Strathdearn having any human life in it. Here are a number of shooting lodges, for these moorlands make grand stalking country. This was and still is Mackintosh country, the chief residing at Moy Hall. The roundness of the summits have resulted in them being generally called 'Cárn' something by the Gaels, Cárn translating as a stony mountain, as opposed to Stob, Sgurr or Beinn, more descriptive of peaks and great bulks.

An extensive area at the head of Glen Killin has been used as a wind farm, as has an area west of Cárn na Saobhaidhe.

10

Height	Name	NGR	OS L	OS E	Ascent
615	Càrn nan Tri-tighearnan	NH 823391	27	E60	
635	Càrn an t-Sean-liathanaich	NH 869320	27/36	E60	
659	Càrn Glas-choire	NH 892291	35/36	E60	
634	Càrn Loisgte	NH 880282	35/36	E60	
627	Càrn a' Choire Mhóir	NH 843291	35	E60	
634	Càrn nam Bain-tighearna	NH 847254	35	E60	
618	Càrn na h-Easgainn	NH 743321	27	E60	
631	Càrn Glac an Eich	NH 694268	26/35	E417	
640	Càrn Odhar Beag	NH 691256	26/35	E417	
642	Aonach Odhar	NH 708230	35	E417	
714	Càrn na Saobhaidh	NH 674241	26/35	E417	
682	Beinn Acha' Bhraghad	NH 657238	26/35	E417	
744	Càrn a' Chaochan-mheadhonach	NH 667217	26/35	E417	
685	Càrn Coire Dhealanaich	NH 708190	35	E417	
807	Beinn Bhreac Mhór	NH 678198	35	E417	
806	Càrn Ghriogair	NH 658199	35	E417	

10

	Height	Name	Grid ref		
☐	675	Coille Mhór	NH 637212	26/35	E417
☐	711	Beinn Bhuidhe	NH 623212	26/35	E417
☐	652	Cárn Leachtar Dhubh	NH 684165	35	E417
☐	802	Cárn Odhar	NH 639179	35	E417
☐	635	Am Bathaich	NH 660139	35	E417
☐	734	Cárn na Dhail-bheag	NH 641137	35	E417
☐	781	Cárn Mhic Iamhair	NH 611145	35	E417
☐	811	△Cárn na Saobhaidhe	NH 600144	35	E417
☐	780	Beinn Bhuraich	NH 582158	35	E417
☐	658	Meall a' Bhuailt	NH 385168	35	E417
☐	689	Beinn Dubhcharaidh	NH 589199	35	E417
☐	728	Cárn Gearresith	NH 628115	35	E417
☐	809	Cárn na Laraiche Maoile	NH 584113	35	E417
☐	797	Doire Mór	NH 576096	35	E56
☐	707	Cárn Liath-bhaid	NH 547112	35	E416
☐	658	Cárn Fliuch-bhaid	NH 551129	35	E416

10

Height	Name	NGR	OS L	OS E	Ascent
	Kinveachy Forest:				
712	Cárn Dearg Mór	NH 862132	35/36	E57	
742	Geal-charn Beag	NH 848144	35	E57	
824	△ Geal-charn Mór	NH 836123	35	E417	
701	Cárn nan Suilean Dubha	NH 820117	35	E417	
652	Cnoc Fraing Beag	NH 808132	35	E417	
745	Cnoc Fraing	NH 807143	35	E417	
618	Cárn Phris Mhóir	NH 807218	35	E417	
636	Cárn Coire na Caorach	NH 802200	35	E417	
622	Sguman Mór	NH 811189	35	E417	
648	Cárn Bad an Daimh	NH 763217	35	E417	
671	Cárn Leachter Beag	NH 765210	35	E417	
750	Cárn Dubh 'Ic an Deoir	NH 774198	35	E417	
740	Cárn Coire na h-Eirghe	NH 756190	35	E417	
672	Cárn na Guaille	NH 783176	35	E417	
635	Creag Dubh Tigh an Aitinn	NH 738180	35	E417	

10

The Mountains of Great Britain

	Name	Grid Ref		
714	Cárn Caol	NH 761161	35	E417
722	Cárn Dubh	NH 765135	35	E417
791	Cárn Coire na h-Easgainn	NH 737136	35	E417
714	Cárn Easgainn Mór	NH 724160	35	E417
773	Cárn Choire Odhair	NH 717142	35	E417
809	Cárn Icean Duibhe	NH 713116	35	E417
778	Cárn Elrig	NH 694127	35	E417
661	Caimhlin Mór	NH 686147	35	E417
814	Calpa Mór	NH 668109	35	E417
812	Cárn Sgulain	NH 697091	35	E56
716	A' Bhuidheanaich Ear - Cnoc an Tiumpain	NH 790088	35	E56
727	A' Bhuidheanaich Iar	NH 773085	35	E56
715	Meall a' Chocaire	NH 760075	35	E56
843	Beinn Bhreac	NH 740069	35	E56
708	Cairn Dulnan	NH 752107	35	E417
878	△ Cárn an Fhreiceadain	NH 725072	35	E56
662	Creag Mhór	NH 734029	35	E56

Height	Name	NGR	OS L	OS E	Ascent
787	Creag Dhubh	NH 728036	35	E56	
749	Cárn Coire na h-Inghinn	NH 720043	35	E56	
766	Geal Charn	NH 703049	35	E56	
853 est	Cárn a' Bhothain Mholaich	NH 706067	35	E56	
756	Creag Dhubh	NH 678972	35	E56	
743	Creag Liath	NH 663007	35	E56	
815	Creag na h-Iolaire	NH 672020	35	E56	
889	Geal Charn	NH 669031	35	E56	
930	▲A' Chailleach	NH 681041	35	E56	
920	▲Cárn Sgulain	NH 683058	35	E56	
911	Meall a' Bhothain	NH 663056	35	E56	
920	Cárn Ballach	NH 643045	35	E56	
942	Cárn Ban	NH 632032	35	E56	
945	▲Cárn Dearg	NH 635024	35	E56	
805	Cárn Macoul	NH 640004	35	E56	
844	Cárn an Leth-choin	NN 623997	35	E56	

10

The Mountains of Great Britain

	Name	Grid Ref		
795	Leacainn Chorrach	NN 598992	35	E56
828	Beinn a' Chrasgain	NN 606981	35	E56
813	Cárn Caol	NN 602968	35	E56
834	Stob Marg na Craige	NN 621973	35	E56
745	Meall na h-Uinneig	NN 627968	35	E56
774	Ileach Bhan	NH 640102	35	E417
825	Cárn Coire na Creiche	NH 623087	35	E56
831	Stob Coire nan Aonach	NH 608075	35	E56
828	Burrach Mór	NH 583083	35	E56
801	Cnoc Ban	NH 576074	35	E55
822	An Staonaig Iar	NH 569055	35	E55
838	An Staonaig Ear	NH 583059	35	E56
873	Cárn Donnachaidh Beag	NH 587038	35	E56
895	Cárn Odhar na Criche	NH 601032	35	E56
875	Cárn Eoghan (Cairn Ewan)	NH 589027	35	E56
713	Cárn na Saobhaidhe	NH 536047	35	E55
688	Cnoc na Chraidhleig	NH 472053	34	E55

10

Height	Name	NGR	OS L	OS E	Ascent
779	Cárn Easgann Bana	NH 485063	34	E55	
718	Cnoc na Choire Odhair	NH 502052	35	E55	
700	Meall nan Ruadhag	NH 526067	35	E55	
767	Cárn Dubh	NH 505084	35	55	
699	Meall nan Aighean Beag	NH 497101	34	E416	
	Glendoe Forest:				
715	Cárn an Dubh Lochain (Cairn Vungie)	NH 452078	34	E55	
729	Cárn Mor	NH 443067	34	E55	
727	Cárn Lochan na Stairne	NH 439062	34	E55	
787 est	Stob Coire Ghlas	NH 443047	34	E55	
786	Stob Coire an t-Seilich	NH 433040	34	E55	
775	Creag an Fhir-eoin	NH 428028	34	E55	
768	Creagan na Cailliche Ard	NH 414025	34	E55	
816	△Cárn a' Chuilinn	NH 416034	34	E66	
	Sherramore Forest:				
887	Beinn Sgiath	NN 566981	35	E55	

10

80

926	▲Geal Charn	NN 562998	35	E55
765	Cárn Fraoich	NH 556013	35	E55
813	Creag an Dearg Lochain	NH 539010	35	E55
844	Leathad Gaothach	NN 523994	35	E55
862	△Meall na h-Aisre	NH 515001	35	E55
756	Sidhean Dubh na Cloiche Baine	NH 504015	35	E55
758	Creag Chomaich	NH 498005	34	E55
736	Cárn Dearg	NN 481985	34	E55
765	Creag Mhór	NN 484973	34	E55
761	Meall Caca	NH 486025	34	E55
870	Druim Mór	NH 470002	34	E55
896	△Gairbeinn	NN 460985	34	E55
Coire Dheirrig Forest:				
876	Geal Charn	NN 444989	34	E55
833	Bac nam Fuaran	NH 436009	34	E55
896	Cnoc a' Coire Dherrig (Corrieyairack Hill)	NN 428998	34	E55
859	Srón a' Bhuirich	NN 422978	34	E55

10

Height	Name	NGR	OS L	OS E	Ascent
879	Cárn a' Choire Shesgnan	NN 412979	34	E55	
884	Cárn Leac	NN 407978	34	E55	
760	Creag a' Chail	NN 403959	34	E55	
622	Cárn Ban	NN 558914	35	E55	
635	Creag Liath	NN 492937	34	E55	
916	Stob Coire Dubh	NN 496917	34	E55	
892	Meall a' Chaorainn Mór	NN 483923	34	E55	
1006	▲Cárn Liath	NN 472903	34	E55	
969	Meall an t-Snaim	NN 459905	34	E55	
848	Creag a' Bhanain	NN 432911	34	E55	
1054	▲Stob Poite Coire Ardair	NN 429889	34/42	E55	
1070	Puist Coire Ardair	NN 436873	34/42	E55	
1028	Meall Coire Choille-rais	NN 433862	34/42	E55	
993	An Cearcallach	NN 423853	34/42	E55	
1128	▲Creag Meagaidh	NN 418875	34/42	E55	
889	Cárn Dearg	NN 410893	34/42	E55	

10

	Height	Name	Grid Ref	Map	Area
☐	817	Meall a' Mheanbh-chruidh	NN 394894	34/41	E55/E400
☐	1043	Beinn Tuath a' Chaorainn	NN 383857	34/41	E55/E400
☐	1052	▲Beinn a' Chaorainn	NN 386851	34/41	E55/E400
☐	915	▲Beinn Teallach	NN 361860	34/41	E400
☐	719	Creag Tharsuinn	NN 365885	34/41	E400
☐	834	△Càrn Dearg	NN 345887	34/41	E400
☐	723	Càrn Bhrunachain	NN 334893	34/41	E400
☐	676	Leana Mhór	NN 317879	34/41	E400
☐	658	Creag Dhubh	NN 323825	34/41	E400
☐	805	Poll-gormack Hill	NN 390980	34	E55/E400
☐	694	Leac nan Uan	NN 372966	34	E400
☐	648	Glas Bheinn	NN 377941	34	E400
☐	679	Càrn Dearg Beag	NN 366942	34	E400
☐	768	△Càrn Dearg	NN 357949	34	E400
☐	702	Meallan Odhar	NN 355962	34	E400
☐	817	△Càrn Dearg	NN 349966	34	E400
☐	791	Glas Charn	NN 354977	34	E400

10

Height	Name	NGR	OS L	OS E	Ascent
☐ 746	Càrn na Larach	NN 334966	34	E400	
☐ 712	Beinn Bhan	NN 326966	34	E400	
☐ 635	Meall a' Chomhlain	NN 322937	34	E400	
☐ 644	Leacann Doire Bainneir	NN 303947	34	E400	
☐ 648	Leitir Fhionnlaigh	NN 274920	34	E400	
☐ 803	△ Beinn Iaruinn	NN 296900	34	E400	
☐ 685	Leana Mhór	NN 285878	34/41	E400	
☐ 654	Cnoc a' Choire Ceirsle	NN 248857	34/41	E400	

10

Knoydart & South Kintail

Knoydart is known as the 'rough bounds', for it forms some of the remotest and wildest country in Britain. No roads lead into it, access being either by foot, along some of the finest paths in the country, or else by boat, landing at the solitary village of Inverie. The paths make grand entrances, either from Kinloch Hourn to Barrisdale, or else from Loch Arkaig through Glen Dessarry. A small passenger ferry leaves Mallaig and can drop one off at Inverie, or further east near to Camusrory, the easiest method of arriving quickly. The three Munro summits (Ladhar Bheinn, Luinne Bheinn and Meall Buidhe) are the most popularly climbed summits in Knoydart, but everywhere on this wild peninsula gives exciting walking, and even the stalkers' paths are of considerable quality.

Part of Knoydart is owned by the Knoydart Foundation, a community-owned trust, which bought 17,500 acres in 1999. The northern side of Ladhar Bheinn (Li and Coire Dhorrcail) has been owned by the John Muir Trust since 1987, which was actually formed in 1983 to try to prevent the Ministry of Defence from buying Knoydart as a bombing range. The peninsula is a National Scenic Area, as is much of South Kintail.

Loch Hourn is one of the country's finest sea lochs, regarded by experts as the nearest Scottish equivalent to a fjord. The mountains rise steeply from either side of the long and narrow channel. Loch Nevis, at the southern end of Knoydart, is similar. Barrisdale bunkhouse makes a good base for Ladhar Bheinn and Meall Buidhe, the former reached by a path into Coire Dhorrcaill, missing the great rocky cliffs at the head of the corrie. From Barrisdale a path climbs to the pass of Mám Barrisdale, from where an ascent of Luinne Bheinn can be made. The way can be continued on to Meall Buidhe, and a descent to Mám Meadail and then west to Inverie, or else north from the summit over the Druim Tor-choire and back to

Mám Barrisdale.

South Kintail is more accessible in that public roads skirt the mountainous hinterland. Glenshiel, and the main road for Skye, bounds the area to the north, and the South Shiel Ridge gives one of the finest long outings in the mountains. The southern part of the range, around Beinn Sgritheall, is often scaled from Arnisdale, once a busy herring fishing centre, reached by a long and twisting byway across Mám Rátagain and through Glenelg. The Corbetts, Beinn na h-Eaglaise and Beinn nan Caorach are fine hills that also rise above Arnisdale

The Saddle is best climbed from Glen Shiel, a pathway ascending the lower slopes of Meallan Odhar giving access to the Forcan ridge, an airy scramble over Sgurr na Forcan to the top. The way can be made south to the Bealach Coire Mhálagain and back up Sgurr na Sgine, another Munro, and the lesser height of Faochag. By continuing on and over Sgurr a' Bhac Chaolais to Bealach Duibh Leac a pathway can be followed back down to the starting point. The same bealach can be used as a start for the South Shiel Ridge, as the range east over Creag nan Damh, Sgurr an Lochain, Maoile an t-Searraich, Maol Chinn-dearg, Aonach Air Chrith, Druim Shionnach and Creag a' Mhaim is known. This gives a long day's walking, with many possible exit routes either down the various northern ridges or into the intervening corries. The northern slopes of the ridge form part of Glenshiel estate. The south side of this range is served by a mountain path that crosses Bealach Duibh Leac and continues past Alltbeithe to Loch Loinne or into Glengarry.

Sgurr a' Mhaoraich lies to the south of this ridge, a solitary Munro which is best climbed from Loch Cuaich (Loch Quoich). A path climbs the Bac nan Cannaichean and so to the top. The way north over Am Báthataich leads to another path back down to Glen Cuaich at Alltbeithe. This forms part of Wester Glenquoich estate.

11

The Mountains of Great Britain

Height	Name	NGR	OS L	OS E	Ascent
759	Creag Bealach na h-Oidhche	NG 836148	33	E413	
974	▲ Beinn Sgritheall	NG 836127	33	E413	
906	Stob Coire Min	NG 845123	33	E413	
805	△ Beinn na h-Eaglaise	NG 854119	33	E413	
774	△ Beinn nan Caorach	NG 871122	33	E413	
618	Beinn Clachach	NG 876106	33	E413	
643	Stob Coire Luachrach	NG 886109	33	E413	
627	Sgurr Mór	NG 879079	33	E413	
713	Druim Fada	NG 894083	33	E413	
709	Sgurr na Laire Brice	NG 890128	33	E413	
662	Druim nan Firean	NG 905126	33	E413	
779	△ Sgurr Mhic Bharraich	NG 917174	33	E413	
680	Sgurr Mheadhonach	NG 910610	33	E413	
683	Torr Coire nan Crogachan	NG 913157	33	E413	
690	Sgurr a' Gharg Gharaidh	NG 916155	33	E413	
680	Sgurr Dheas a' Gharg Gharaidh	NG 917151	33	E413	

11

The Mountains of Great Britain

Height	Name	NGR	OS L	OS E	Ascent
919	Sgurr Leac nan Each	NG 918133	33	E413	
939	Spidean Dhomhnuill Bhric	NG 922129	33	E413	
1011	▲ The Saddle	NG 934131	33	E414	
644	Biod an Fhithich	NG 951147	33	E413/E414	
909	Faochag	NG 954123	33	E414	
946	▲ Sgurr na Sgine	NG 946114	33	E414	
885	Sgurr a' Bhac Chaolais	NG 958110	33	E414	
815	Torr an Damhain	NG 964101	33	E414	
826 est	Cárn a' Choire Sgoireadail	NG 965099	33	E414	
885	△ Buidhe Bheinn Ear	NG 963090	33	E414	
879	Buidhe Bheinn	NG 957087	33	E414	
755	Sgurr a' Chuilinn	NG 982121	33	E414	
918	▲ Creag nan Damh	NG 983112	33	E414	
896	Sgurr Beag	NG 997109	33	E414	
1004	▲ Sgurr an Lochain	NH 005104	33	E414	
1010	▲ Sgurr an Doire Leathain	NH 015099	33	E414	

11

The Mountains of Great Britain

	Height	Name	Grid Ref		
☐	902	Sgurr Coire na Feinne	NH 068092	33	E414
☐	981	▲Maol Chinn-dearg	NH 032088	33	E414
☐	1021	▲Aonach Air Chrith	NH 051083	33	E414
☐	938	Stob Coire an t-Slugain	NH 063082	33	E414
☐	987	▲Druim Shionnach	NH 074084	33	E414
☐	947	▲Creag a' Mhaim	NH 088078	33	E414
☐	906	Sgurr Thionail	NG 985089	33	E414
☐	899	Am Bathaich	NG 988075	33	E414
☐	1027	▲Sgurr a' Mhaoraich	NG 984065	33	E414
☐	891	Sgurr Coire nan Eiricheallach	NG 999061	33	E414
		Barrisdale Forest:			
☐	841	Sgurr a' Chlaideimh	NG 952031	33	E414
☐	835 est	Stob Coire Shubh	NG 942035	33	E414
☐	895	△Sgurr nan Eugallt	NG 931044	33	E414
☐	738	Sgurr Dubh	NG 941055	33	E414
☐	666	Càrn an Lochain	NG 903052	33	E413
☐	667	Meall nan Eun	NG 903052	33	E413

11

Height	Name	NGR	OS L	OS E	Ascent
622	An Caisteal	NG 893043	33	E413	
700	Slat Bheinn	NG 910027	33	E413	
913	△Sgurr a' Choire-bheithe	NG 896016	33	E398/E413	
757 est	Druim Chosaidh	NG 917007	33	E398/E413	
777	Sgurr Airigh na Beinne	NG 924007	33	E398/E413	
648 est	Meall an Spardain	NG 941007	33	E398/E414	
939	▲Luinne Bheinn	NG 869008	33	E398/E413	
839	Druim Leac a' Shith	NM 868993	33/40	E398/E413	
946	▲Meall Buidhe	NM 849989	33/40	E398/E413	
793	Sgurr Sgeithe	NM 862983	33/40	E398/E413	
826	An t-Uiriollach	NM 840991	33/40	E398/E413	
718	Meall Bhasiter	NM 946972	33/40	E398/E413	
855	△Beinn Bhuidhe	NM 822967	33/40	E398	
787	Sgurr Coire nan Gobhar	NM 804970	33/40	E398/E413	
796	△Sgurr Coire Choinneachain	NG 791011	33	E398/E413	
661	Stob an Uillt-fhearna	NG 807018	33	E398/E413	

11

	Height	Name	Grid Ref	Sheet	Explorer
☐	620	Càrn a' Mhaim Suidheig	NG 822018	33	E398/E413
☐	758	Aonach Sgoilte Beag	NG 831022	33	E413
☐	849	Aonach Sgoilte	NG 840027	33	E413
☐	840	Stob a' Chearcaill	NG 846029	33	E413
☐	858	Stob Choire Dhorrcail	NG 833033	33	E413
☐	1020	▲ Ladhar Bheinn	NG 824040	33	E413
☐	960	Stob a' Choire Odhair	NG 830043	33	E413
☐	692	Stob Coire Each	NG 824058	33	E413
☐	668	Mullach Li	NG 817064	33	E413
☐	785	△ Beinn na Caillich	NG 796067	33	E413
☐	686	Meall Coire an t-Searraich	NG 781056	33	E413

11

Loinne, Airceag & Morar

This section covers the mountainous region south of Glen Morriston, west of the Loch Lochy section of the Great Glen, north of lochs Eils and Eilt, and westwards to the adjoining Knoydart peninsula. Perhaps the best known peak in the area, to mountain lovers, is the distinctive cone of Sgurr na Ciche, rising at the head of Loch Nevis. A number of great lochs occupy the glen bottoms here, many enlarged by the erection of dams, such as lochs Cluanie (1957), Cuaich and Airceag (or Arkaig) and Loinne (1956). The creation of Loch Loinne meant that the Skye road had to be rerouted—formerly it went by way of Tomdoun and the south end of Loch Cluanie. Today it begins climbing up the north side of Loch Garry, swings back to Bunloinne, then follows the northern shores of Loch Cluanie.

The peaks of Glengarry Forest give excellent walking, Meall na Teanga and Srón a Choire Ghairbh rising to Munro status, but Beinn Tee is perhaps one of the finest in this group. Meall na Teanga is best climbed from Clunes, near Burnarkaig, crossing the shoulder of Leach Chorrach. Srón a' Choire Ghairbh and Beinn Tee make a fine circular walk from Kilfinnan, at the head of Loch Lochy.

Gleouraich and Spidean Mialach make an interesting ridge-walk. From Loch Cuaich a path ascends Gleouraich's western finger. The ridge eastward is followed over the east top (Gleouraich Ear) to Spidean Mialach, from where a descent is made to a path from Coire Mheil back to the starting point.

The wildest summits in this section are those of Sgurr Mór, Sgurr na Ciche and the other heights around Glen Dessarry. A road follows the northern shores of Loch Airceag to Strathan, where cars can be left. A forest road passes A' Chuil bothy in Glen Dessarry, continuing as a path to Sourlies bothy. A steep ascent can be made to Garbh Chioch Mór and thence to Sgurr na Ciche. From the same path Sgurr

12

nan Coireachan can be climbed, and a way made over An Eag and Sgurr Beag to Sgurr Mór. From the southern base of this mountain a pathway returns through a glen to Glendessarry House. Another possibility is to drop to Kinbreack bothy, from where another path returns to Strathan.

Gaor Bheinn (also known as Gulvain) is often climbed from Gleann Fionnlighe, a track then path leading up the glen to the southern top and then to the main summit itself, with rugged Coire Sgreamhach below. Sgurr Thuilm and Sgurr nan Coireachan make a fine round trip from Glen Finnan, passing through the famous curved concrete Glenfinnan Viaduct. A bothy at Corryhully is a useful stop-over point.

The lower mountains to the west, though they fail to attract Munroists, give excellent wild walking. The ridge from Tarbet on Loch Nevis east to the head of Glen Dessarry is one of the most remote in the area. Six summits on this are in excess of 2,000 feet, the highest being Sgurr na h-Aide, listed in Corbett's Tables. This area forms North Morar deer forest.

There are a few notable Corbetts in this section—Beinn Bhán being prominent between Glen Loy and Loch Airceag. Streap is a distinguished summit, almost reaching Munro status.

Loch Airceag (or Arkaig) is one of the most attractive lochs hereabouts, reached through the Mile Dorcha, famed in Jacobite story. Achnacarry is the seat of Cameron of Lochiel, chief of the clan, and thousands of acres of countryside around the loch are his. The estate extends south towards Loch Eil.

Loch Morar, the deepest fresh-water loch in Europe, is surrounded by 2,000 feet mountains only at its eastern end, a difficult place to reach. This remoteness was not lost on Bonnie Prince Charlie, who passed this way on his escape from the Battle of Culloden, the roughness of the walking following his defeat in battle taking his morale to an all-time low. A cairn on the shore of Loch nan Uamh marks where he first landed on his return from France. Above Loch Beoraid is a cave that he used to hide in.

12

The Mountains of Great Britain

Height	Name	NGR	OS L	OS E	Ascent
☐ 715	Mám a' Chroisg	NH 267080	34	E400	
☐ 660	Beinn an Eoin	NH 242090	34	E415	
☐ 788	△Meall Dubh	NH 244078	34	E415	
☐ 680	Cárn Dearg	NH 228069	34	E415	
	Coille Bun-loinne:				
☐ 750	Druim na Garbh-leitir	NH 161082	34	E415	
☐ 775	△Beinn Loinne	NH 151078	34	E414/E415	
☐ 790	Druim nan Cnamh	NH 131077	34	E414	
	East Glenquoich Forest:				
☐ 1035	▲Gleouraich	NH 039052	33	E414	
☐ 1008	Creag Coire na Fiar Bhealaich	NH 047051	33	E414	
☐ 614	Cárn Loch Fearna	NH 053033	33	E399/E400	
☐ 996	▲Spidean Mialach	NH 066043	33	E400	
☐ 735	Creag Dhubh	NH 086040	33	E400	
	Glengarry Forest:				
☐ 904	△Beinn Tee	NN 241972	34	E400	

12

The Mountains of Great Britain

☐	901	Meall a' Choire Ghlais	NN 219957	34	E400
☐	937	▲ Srón a' Choire Ghairbh	NN 223945	34	E400
☐	888	Sean Mheall	NN 242947	34	E400
☐	693	Meall nan Dearcag	NN 257953	34	E400
☐	839	Meall Dubh	NN 229932	34	E400
☐	918	▲ Meall na Teanga	NN 220924	34	E400
☐	907	Meall Coire Lochain	NN 215920	34	E400
☐	761	Meall an Tagraidh (Meall an t-Sagairt)	NN 194941	34	E399/E400
☐	838	△ Meall na h-Eilde	NN 185946	34	E399/E400
☐	826	Meall Coire na Saobhaidh	NN 174952	34	E399/E400
☐	660	Meall Tarsuinn	NN 168960	34	E399
☐	732	Glas Bheinn	NN 172919	34	E399/E400
☐	804	△ Geal Charn	NN 156943	34	E399
☐	616	Beinn Chraoibh	NN 142926	34	E399
☐	749	Sgurr Choinnich	NN 127949	34	E399
☐	656	Meall Blair	NN 079951	33	E399
☐	880	△ Sgurr Mhurlagain	NN 012944	33	E399

Height	Name	NGR	OS L	OS E	Ascent
858	△Fraoch Bheinn	NM 986940	33/40	E398/E399	
822	Druim a' Chuirn	NM 959951	33/40	E398	
835	△Sgurr Cos na Breachd-laoidh	NM 948947	33/40	E398	
873	An Eag	NM 943959	33/40	E398	
890	Sgurr Beag	NM 959971	33/40	E398	
1003	▲Sgurr Mór	NM 965980	33/40	E398	
901	△Sgurr an Fhuarain	NM 987980	33/40	E398/E399	
919	▲Gairich	NN 026996	33	E399	
953	▲Sgurr nan Coireachan	NM 932958	33/40	E398	
695	An t-Sàil	NM 941979	33/40	E398	
1013	▲Garbh Chioch Mhór	NM 909961	33/40	E398	
1040	▲Sgurr na Cìche	NM 902967	33/40	E398	
760	Càrn a' Choire na Cuairtich	NM 895962	33/40	E398	
740	Meall a' Choire Dhuibh	NM 919981	33/40	E398/E413	
717	Càrn Dearg	NM 907983	33/40	E398/E413	
887	△Beinn an Aodainn	NM 899986	33/40	E398/E413	

12

Height	Name	Grid Ref		
612	Sgurr Mór	NM 826921	33/40	E398
728	Sgurr Breac	NM 846924	33/40	E398
747	Sgurr nam Meirleach	NM 864930	33/40	E398
859	Sgurr na h-Aide	NM 882932	33/40	E398
867	△Bidein a' Chabhair	NM 889931	33/40	E398
674	Meall na Sroine	NM 906939	33/40	E398
829	△Cárn Mór	NM 903909	33/40	E398
616	Stob Coire an Eich	NM 925901	33/40	E398
710	Meith Bheinn	NM 821872	40	E398
718	An Stac	NM 866889	40	E398
817	Sgurr an Ursainn	NM 875869	40	E398
896	Beinn Gharbh	NM 882877	40	E398
633	Glas-charn	NM 846837	40	E398
796	△Sgurr an Utha	NM 884840	40	E398
790	Fraoch-bheinn	NM 895837	40	E398
749	Sgurr an Fhuarain Duibh	NM 901859	40	E398
852	Sgurr a' Choire Riabhaich	NM 908871	40	E398

12

Height	Name		NGR	OS L	OS E	Ascent								
956	▲ Sgurr nan Coireachan		NM 903880	40	E398									
826	Meall an Tarmachain		NM 911883	40	E398									
825	Beinn Gharbh		NM 923881	40	E398									
963	▲ Sgurr Thuilm		NM 939880	40	E398									
810	Beinn an Tuim		NM 929835	40	E398									
844	Meall an Uillt Chaoil		NM 932843	40	E398									
887	Stob Coire nan Cearc		NM 937852	40	E398									
909	△ Streap		NM 946864	40	E398									
898	Streap Comhlaidh		NM 952861	40	E398									
691	Na h-Uamhachan		NM 967843	40	E398									
765	△ Bràigh nan Uamhachan		NM 975867	40	E398									
961	Gaor Bheinn (South West Top)		NM 997865	40	E398/E399									
987	▲ Gaor Bheinn (Gulvain)		NN 002876	41	E399									
727	Mullach Coire nan Geur-oirean		NN 049893	41	E399									
649	Beinn an t-Sneachda		NM 985815	40	E398/E399									
663	Aodann Chleireig		NM 994826	40	E398/E399									

12

681	Meall Onfhaidh	NN 011981	41	E399
774	△ Meall a' Phubuill	NN 029854	41	E399
747	Monadh Mór	NN 041854	41	E399
698	Druim Gleann Laoigh	NN 062853	41	E399
729	Druim Fada	NN 061822	41	E392/E399
744	Stob a' Ghrianain	NN 087824	41	E392/E399
796	△ Beinn Bhàn	NN 141857	41	E399

12

Ardghabhar & Muideart

Ardghabhar and Muideart (or Ardgour and Moidart as they are better known) are separated by the long stretch of Loch Shiel. Muideart was part of Inverness-shire, whereas Ardghabhar was part of Argyll, but now both fall within Highland region. Although no summit rises above the Munroist's magical 3,000 foot contour, the peaks are none the worse for that, Rois Bhenn and Garbh Bheinn being two of the grandest peaks in the west. Both are within easy distance of public roads.

Rois Bheinn has two summits, a straightforward climb up its western shoulder from near Roshven farm. An attractive ridge, with the summits of Sgurr na Ba Glaise and Druim Fiaclach extends further east. Transport at both ends is the best way of climbing these mountains. Beinn Odhar Mhór is reached from Glenfinnan village, following the Allt na h-Aire to Lochan nan Sleubhaich and thence up the north-east face of the mountain. The next summit, Beinn Odhar Bheag, is actually higher, despite the Gaelic names being the other way round, and it is listed as a Corbett.

Garbh Bheinn has many rock climbs in its north-eastern corrie. The casual walker will find the best way to the twin tops is to follow the Srón a Gharbh Choire Bhig, up which a feint path can be found. A better day's walking can be had by traversing Coire an Iubhair, preferably in an anti-clockwise direction, starting with Sgorr Mhic Eacharna. From Beinn Bheag the ridge should be followed westwards before dropping to Loch Coire an Iubhair, from where Garbh Bheinn is a short, but steep, climb.

Loch Shiel is one of the most attractive lochs of the western highlands, its mountainous sides covered in natural woodlands. At the head of the loch stands the Glenfinnan Monument, marking where the standard of Prince Charles Edward Stewart was erected on 19 August 1715. This is owned by the National Trust for Scotland, who operate a visitor centre here. On the west side of the loch is the setting of *A Last Wild Place*, a classic nature book written by Mike Tomkies.

The more distant summits are less frequented, even although good tracks and paths make their way through some of the glens. Resourie bothy, at the head of Glen Hurich, makes a useful stop-over point for those heading deeper into the hills. The highest peak in this area is Sgurr Dhomhnuill, a remote summit no matter from where it is climbed. The eastern shore of Loch Shiel has a track as far as Polloch, where it becomes public and crosses the lesser hills of Sunart to Strontian. Here the element Strontium was first discovered in 1791, getting its name from the village. This was an important lead-mining centre for over 150 years, the remnants of the excavations still visible.

Beinn Resipol is a Corbett which rises alone to the north-west of Strontian. An ascent can be made from that village, a path ascending the northern slopes of Beinn a' Chaorainn giving access to the upper slopes. A shorter route can be made from Resipol or Ardery.

South of the Loch Sunart-Glen Tarbet gap one is in Morvern. The highest peaks are in the Kingairloch section here, arranged in a great arc around Loch a' Choire. Beinn Mheadhoin has four interesting ridges descending from its summit. These mountains are made of granite rock, and in Glen Sanda is a massive quarry extracting this, causing a blight on the landscape. The quarry, located between Meall na Fidhle and Meann na h-Easaiche, was established in 1986 and the first shipload of granite was exported to the United States. Apparently, there are enough reserves to last one hundred years. The horseshoe round Glen Galmadale makes a grand traverse. From Glengalmadale the Druim na Maodalaich is followed over Meall nan Each to Maol Odhar, Creach Bheinn (the highest summit in Kingairloch, and listed by Corbett) and south over Fuar Bheinn (another Corbett) and Beinn na Cille to Camasnacroise or Ceann Ghearr Loch.

13

Height	Name	NGR	OS L	OS E	Ascent
870	Beinn Odhar Mhór	NM 851791	40	E391/E398	
882	△ Beinn Odhar Bheag	NM 846779	40	E391	
773	Beinn a' Chaorainn	NM 839773	40	E391	
783	△ Beinn Mhic Cedidh	NM 828788	40	E391	
665	Diollaid Bheag	NM 812802	40	E391/E398	
751	Diollaid Mhór	NM 807795	40	E391/E398	
787	Beinn Coire nan Gall	NM 792798	40	E390/E398	
869	Druim Fiaclach	NM 791792	40	E390/E398	
831 est	An t-Slat-bheinn	NM 778778	40	E390	
874	△ Sgurr na Ba Glaise	NM 770777	40	E390	
814	△ An Stac	NM 763793	40	E390/E398	
882	△ Rois Bheinn	NM 756778	40	E390	
878	Rois Bheinn Iar	NM 749778	40	E390	
713	Sgurr Dhomhnuill Mór	NM 740759	40	E390	
666	Beinn Gaire	NM 781749	40	E390	
663	Croit Bheinn	NM 811774	40	E391	

13

845	△Beinn Resipol	NM 766655	40	E390
619	Beinn an Albannaich	NM 763649	40	E390
771	△Stob Coire a' Chearcaill	NN 107727	41	E391
722	Sgúrr an Iubhair	NN 001720	41	E391
636	Glas Bheinn	NM 938758	40	E391
723	Meall nan Damh	NM 919745	40	E391
722	Meall Iar nan Damh	NM 911747	40	E391
668	Sgórr nan Cearc	NM 898772	40	E391
643 est	Meall Doire na Mnatha	NM 896770	40	E391
775	△Sgórr Craobh a' Chaorainn	NM 896758	40	E391
634	Meall a' Choire Chruinn	NM 879762	40	E391
849	△Sgurr Ghiubhsachain	NM 875751	40	E391
755	Meall nan Creag Leac	NM 861747	40	E391
756	Sgórr an Tarmachain	NM 839714	40	E391
701	Teanga Chorrach	NM 865723	40	E391
770	△An Sgriodain (Druim Tarsuinn)	NM 874727	40	E391
759	Meall Mór	NM 887728	40	E391

13

Height	Name	NGR	OS L	OS E	Ascent
716	Stob a' Chuir	NM 903717	40	E391	
706	Srón a' Choire Leith Mhóir	NM 910717	40	E391	
721	Stob MhicBheathain	NM 914713	40	E391	
786	△ Cárn na Nathrach	NM 887699	40	E391	
803	Druim Garbh	NM 882683	40	E391	
888	△ Sgurr Dhomhnuill Mór	NM 889 679	40	E391	
766	Sgurr na h-Ighinn	NM 887670	40	E391	
761	Sgurr a' Chaorainn	NM 895662	40	E391	
624	Sgurr na Laire	NM 898655	40	E391	
762	△ Beinn na h-Uamha	NM 917664	40	E391	
734	Meall Dearg Choire nam Muc	NM 979657	40	E384/E391	
730	Sgurr na h-Eanchainne	NM 997658	40	E391	
701	Sgurr nan Cnamh	NM 887643	40	E391	
627	Druim Min	NM 892634	40	E391	
736	Beinn Bheag	NM 914636	40	E391	
650	Sgórr Mhic Eacharna	NM 929630	40	E391	

13

	Height	Name	Grid Ref		Map
☐	687	Meall a' Chuilinn	NM 892613	40	E391
☐	773	Garbh Bheinn Iar	NM 892618	40	E391
☐	885	△ Garbh Bheinn	NM 903622	40	E391
☐	823	Srón a' Gharbh Choire Mhóir	NM 907617	40	E391
☐	794	Maol Odhar	NM 881579	49	E376/E383
☐	853	△ Creach Bheinn	NM 871576	49	E376/E383
☐	766	△ Fuar Bheinn	NM 853563	49	E376/E383
☐	652	Beinn na Cille	NM 854542	49	E376/E383
☐	626	Cárn na Tri Sgirean	NM 836575	49	E383
☐	642	Glas Bheinn	NM 834566	49	E383
☐	739	Beinn Mheadhoin	NM 799514	49	E383
☐	702	Meall na Fidhle	NM 802507	49	E383

13

Killiechonate & Mamore Forests

The highest mountain in Great Britain, Ben Nevis, is located within this section, rising as it does steeply above Fort William. The summit is 4,413 feet above sea level, an energetic climb from Glen Nevis being the most popular route, which attracts over 100,000 climbers per annum. Being the highest mountain, between 1883 and 1904 an observatory was located on the summit, manned by meteorological staff who collected data, such as a mean annual rainfall of 157 inches, though this peaked at 240 inches one year. It wasn't until 1847 that Ben Nevis was confirmed by the Ordnance Survey map-makers as being the tallest mountain in Britain, previously Ben MacDuibh being thought to hold this honour. The two Cárn Deargs are subsidiary summits of Ben Nevis, and both are in excess of 1,000 metres. To the east is the Cárn Mór Dearg arete, which leads to Cárn Mór Dearg itself, another lofty summit at 4,003 feet.

To the east of Ben Nevis is the Killiechonate Forest, with the two summits of Aonach Mór and Aonach Beag at the western end. The northern slopes of Aonach Mór have been developed as a ski centre, known as Nevis Range, with a gondola carrying visitors, skiers and mountain cyclists to a spot just above Sgurr Finniosgaig. A range of high mountains strikes east from here, known colloquially as the Grey Corries from the series of corries on the northern slopes. Three of the mountains are Munros—Sgurr Choinnich Mór, Stob Coire an Laoigh and Stob Choire Claurigh. Between, and to all sides of, these peaks are lesser summits that merit mountain status listing, and which are Munro tops. Just off the Grey Corries ridge is Stob Bán, another Munro.

East of the Grey Corries are two Munros that are linked by a high ridge—Stob Coire Easain and Stob a' Choire Mheadhoin. Both are in excess of 1,000 metres, rising above Loch Treig. North of here are Cruach Innse and Sgurr Innse, two fairly solitary mountains that merit Corbett listing.

At the head of Loch Treig are a number of fair-sized hills that stretch south-west towards the Blackwater Reservoir. Glas Bheinn is a Corbett.

The Mamore Forest contains a long ridge of mountains, the ascent of all of the Munro summits being a lengthy day's challenge. A number of the tops are in excess of 1,000 metres, making this part of the Highlands one of the highest. North of Loch Eilde Mór rises Sgúrr Eilde Mór, the eastmost Munro. Binnein Beag is a conical Munro, its larger brother, Binnein Mór, rising to its south-west. Na Gruagaichean has two summits, the south-east one being the taller, and meriting listing by Munro.

Circling Coire a' Mhail are four Munros, plus three other Munro tops. An Gearanach is at the eastern side of the corrie, and working clockwise are the tops of An Garbhanach, Stob Coire a' Chairn, Am Bodach, Sgúrr an Iubhair, Stob Choire a' Mhail and Sgúrr a' Mháim. The last-named is the highest at 3,601 feet.

Stob Bán is a spectacular peak with a rocky east face. It fails to make the 1,000 metre mark by just one metre. Mullach nan Coirean is the westmost Munro in this range.

To the south and west of the Lairig Mór are a few summits that rise above the 2,000 feet mark. Beinn na Gucaig and Tom Meadhoin just make mountain status by a few feet, whereas Mám na Gualainn and Beinn na Caillich are much taller, the former being a Corbett.

Some of the periphery of this section is owned by Scottish Land and Forestry, such as the Leanachan, Nevis and Glen Righ forests. Much of Ben Nevis and Aonach Beag is owned by the John Muir Trust, which acquired it in 2000. The vast bulk of the section is the property of British Alcan Aluminium, successor to the British Aluminium Company, which built the dams from 1901 onwards.

14

Height	Name	NGR	OS L	OS E	Ascent
711	Meall an t-Suidhe	NN 139730	41	E392	
1221	Cárn Dearg	NN 159719	41	E392	
1345	▲ Ben Nevis (Beinn Nibheis)	NN 167713	41	E392	
1020	Cárn Dearg	NN 155701	41	E392	
698	Meall Cumhann	NN 178697	41	E392	
1220	▲ Cárn Mór Dearg	NN 178722	41	E392	
	Killiechonate Forest:				
663	Sgurr Finniosgaig	NN 189762	41	E392	
1221	▲ Aonach Mór	NN 193730	41	E392	
1068	Stob an Cul Choire	NN 203731	41	E392	
918	Tom na Sroine	NN 208747	41	E392	
1234	▲ Aonach Beag	NN 196714	41	E392	
963	Sgurr a' Bhuic	NN 204702	41	E392	
963	Sgurr Choinnich Beag	NN 220710	41	E392	
1094	▲ Sgurr Choinnich Mór	NN 227715	41	E392	
1080	Stob Coire Easain	NN 234726	41	E392	

14

The Mountains of Great Britain

	Height	Name	Grid ref	Map	Sheet
☐	1116	▲ Stob Coire an Laoigh	NN 240728	41	E392
☐	1106	Caisteal	NN 245729	41	E392
☐	1177	▲ Stob Choire Claurigh	NN 262738	41	E392
☐	1123	Stob Coire na Ceannain	NN 268746	41	E392
☐	958	Stob Coire Gaibhre	NN 261757	41	E392
☐	724	Beinn Chlianaig	NN 293782	41	E392
☐	742	Cnap Cruinn	NN 303775	41	E392
☐	857	△ Cruach Innse	NN 280763	41	E392
☐	809	△ Sgurr Innse	NN 290748	41	E392
☐	1105	▲ Stob a' Choire Mheadhoin	NN 317737	41	E392
☐	1115	▲ Stob Coire Easain	NN 308730	41	E392
☐	723	Creagan a' Chaise	NN 311708	41	E385
☐	621	Creag Ghuanach	NN 300690	41	E385
☐	721	Meall Mór	NN 281706	41	E385/E392
☐	841	Meall a' Bhuirich	NN 254706	41	E392
☐	977	▲ Stob Ban	NN 266746	41	E392
☐	633	Meall a' Bhainne	NN 306664	41	E385

14

Height	Name	NGR	OS L	OS E	Ascent
646	Beinn na Cloiche	NN 285649	41	E385/E392	
792	△ Glas Bheinn	NN 259641	41	E392	
	Mamore Forest:				
1010	▲ Sgurr Eilde Mór	NN 231658	41	E392	
943	▲ Binnein Beag	NN 222677	41	E392	
1130	▲ Binnein Mór	NN 212663	41	E392	
1056	▲ Na Gruagaichean	NN 203652	41	E392	
1041	Na Gruagaichean Tuath	NN 202604	41	E392	
982	▲ An Gearanach	NN 188670	41	E392	
981	▲ Stob Coire a' Chairn	NN 185661	41	E392	
909	Stob Deas Coire a' Chairn	NN 181657	41	E392	
1032	▲ Am Bodach	NN 177651	41	E392	
1001	Sgórr an Iubhair	NN 165655	41	E392	
990	Stob Choire a' Mhail	NN 163660	41	E392	
1099	▲ Sgurr a' Mhaim	NN 164668	41	E392	
999	▲ Stob Ban	NN 148654	41	E392	

	Height	Name	Grid Ref		Code
☐	912 est	Druim nan Coirean	NN 138656	41	E392
☐	917	Mullach Ear nan Coirean	NN 131654	41	E392
☐	939	▲Mullach nan Coirean	NN 122662	41	E392
☐	910	Meall a' Chaorainn	NN 114657	41	E392
☐	764	Beinn na Caillich	NN 141628	41	E384/E392
☐	755	Cárn Loch an Sgoir	NN 122627	41	E384/E392
☐	796	△Mám na Gualainn	NN 115625	41	E384/E392

Coille Gleann-righ

	Height	Name	Grid Ref		Code
☐	621	Tom Meadhoin	NN 987621	41	E384/E392
☐	615	Creag Bhreac	NN 978618	41	E384/E392
☐	616	Beinn na Gucaig	NN 063653	41	E384/E392

14

Ben-alder Forest & Srath Ossian

The remoteness of the mountainous countryside that forms this section was known to Bonnie Prince Charlie who spent some time hiding here in September 1746, following the Battle of Culloden. Prince Charlie's Cave, or Cage, as it is properly known, overlooks Alder Bay on Loch Ericht, on the lower slopes of the Ben Alder massif. At 3,765 feet, Ben Alder is one of the highest mountains in the country, one that requires considerable walking to reach. Its flat granite top is enlivened by the sheer drops to the corries below. Many ascents are made from the south, from Bridge of Ericht at Loch Rannoch, or else from the west, the remote Corrour Railway Station allowing access to the walker. Routes from the north tend to be much longer. On a clear day views as far south to the Lomond Hills in Fife are possible.

On the east side of Bealach Beithe is Beinn Bheóil, another Munro, and Srón Coire na h-Iolaire, the former in excess of 1,000 metres. On the opposite side of Bealach Cumhann is Beinn a' Chumhainn, and a trio of Munros, arranged along the boundary between the Rannoch and Corrour forests. Although not as tall as Ben Alder, the summits of Sgór Gaibhre, Sgór Choinnich and Cárn Dearg are distinctive peaks, worth climbing. To the south, within Rannoch Forest, is Meall na Meoig, sometimes referred to as Beinn Pharlagain, or its northern top, which is a Corbett.

North of the two Munros lies Loch Ossian, a remote loch with the modern Corrour Lodge at its eastern end and a simple youth hostel at its west. The forestry here was pioneered by Sir John Stirling Maxwell. Unlike most of the lochs in the surrounding hills, Loch Ossian is totally natural. Beyond the railway line rises Leum Uilleim, a Corbett that rises to almost Munro height. A round trip of just two hours can be made from the railway station at Corrour to the summit and back. The West Highland Railway was constructed in 1894.

On the north side of Loch Ossian rises Beinn na Lap, a Munro, and further north again is Chno Dearg, a substantial summit at 3,433 feet. Creagan Coire nan Cnámh on the east side of Meall Garbh, has some rock-climbs. Forming a ring around Coire an Lochain rise the two Stob Coire Sgriodain summits, the northern one of which is a Munro. Ascents of these mountains are usually made from Fersit, near to Tulloch Station, to the north.

Loch Ghuilbinn and Srath Ossian form a large cleft that separates the mountains to the east, a few of which rise over 1,000 metres. Beinn Eibhinn, Aonach Beag, Geal Charn and Cárn Dearg are four Munros that form the topmost part of this ridge, one that makes an excellent adventure, miles from public roads, and with mountain tops of character, deep corries and little corrie-lochans. Just off this ridge lies Sgór Iutharn, one of the most pointed summits when viewed from the east.

Three more Munros are found north of An Lairig, two of them, Beinn a' Chlachair and Geal Charn, being in excess of 1,000 metres, Creag Pitridh being a minnow at 924 metres. Beyond the two Lochan na h-Earba rise two mountain tops with similar names—Binnein Shuas and Binnein Shios. Both rise above the southern shores of Loch Laggan.

East of the River Pattack is a group of hills of which only The Fara is of any significance. It is just eleven feet short of meriting inclusion as a Munro. However, its summit is rather bland, apart from the large cairn, compared with others in the district.

Most of this section is covered by three private sporting estates, the shooting lodges being considerable houses, namely Ben Alder, Corrour and Ardverikie. The last-named is the oldest building, a baronial mansion, and it was almost considered as a highland retreat of Queen Victoria, before she and Albert bought Balmoral. It has since gained some fame from television appearances.

Height	Name	NGR	OS L	OS E	Ascent
658	Meall nan Eagan	NN 597874	42	E50	
649	Creag nan Adhaircean	NN 590863	42	E50	
911	△ The Fara	NN 598843	42	E50	
904	The Fara (South Top)	NN 595829	42	E50	
901	Leacann na Sguabaich	NN 586820	42	E50	
897	Meall Cruaidh	NN 578809	42	E50	
674	Beinn Eilde	NN 563850	42	E50	
	Ardverikie Forest:				
667	Binnein Shios	NN 492857	42	E50	
747	Binnein Shuas	NN 462827	42	E50/E55	
643	Creag a' Chuir	NN 504848	42	E50	
715	Creag a' Mhaigh	NN 500837	42	E50	
842	Srón nan Tarmachan	NN 509832	42	E50	
863	Meall Buidhe	NN 511825	42	E50	
802	Meall na Brachdlach	NN 520193	42	E50	
1049	▲ Geal Charn (Mullach Coire an Iubhair)	NN 504812	42	E50	

	Height	Name	Grid ref		Road
☐	924	▲ Creag Pitridh (Creag Peathraich)	NN 488813	42	E50/E55
☐	1087	▲ Beinn a' Chlachair	NN 471781	42	E50
☐	618	Meall Cos Charnan	NN 433776	42	E50
		Ben-alder Forest:			
☐	1034	▲ Cárn Dearg	NN 503764	42	E50
☐	925	Diollaid a' Chairn	NN 487758	42	E50
☐	1028	Sgór Iutharn	NN 490743	42	E50
☐	1132	▲ Geal Charn	NN 470746	42	E50
☐	1116	▲ Aonach Beag	NN 458742	42	E50
☐	1102	▲ Beinn Eibhinn	NN 449733	42	E50
☐	921	Mullach Coire nan Nead	NN 430734	42	E50
☐	1148	▲ Ben Alder (Beinn All-dobhar)	NN 496719	42	E50/E385
☐	956	Srón Coire na h-Iolaire	NN 512704	42	E50/E385
☐	1019	▲ Beinn Bheoil	NN 517717	42	E50/E385
☐	815	Meall Chaorach	NN 383757	41	E50
☐	1046	▲ Chno Dearg	NN 377741	41	E50
☐	958	Stob Coire Sgriodain Deas	NN 359739	41	E50

Height	Name	NGR	OS L	OS E		Ascent
979	▲ Stob Coire Sgriodain	NN 357743	41	E50		
976	Meall Garbh	NN 372728	41	E50		
857	Garbh-bheinn	NN 358712	41	E50		
704	Meall Glas-uaine Mór	NN 405722	42	E50		
935	▲ Beinn na Lap	NN 377696	41	E50/E385		
876	Beinn a' Bhric	NN 318642	41	E385		
909	△ Leum Uilleim	NN 331641	41	E385		
849	Srón Uilleam	NN 333633	41	E385		
	Coir'-odhair Forest:					
903	Beinn a' Chumhainn	NN 463710	42	E50/E385		
865	Meall a' Bhealaich	NN 453694	42	E50/E385		
929	Sgór Choinnich	NN 443683	42	E50/E385		
955	▲ Sgór Ghaibhre	NN 444674	42	E385		
941	▲ Cárn Dearg	NN 418661	42	E385		
861	Meall nam Fiadh	NN 421652	42	E385		
868	△ Meall na Meoig	NN 448642	42	E385		

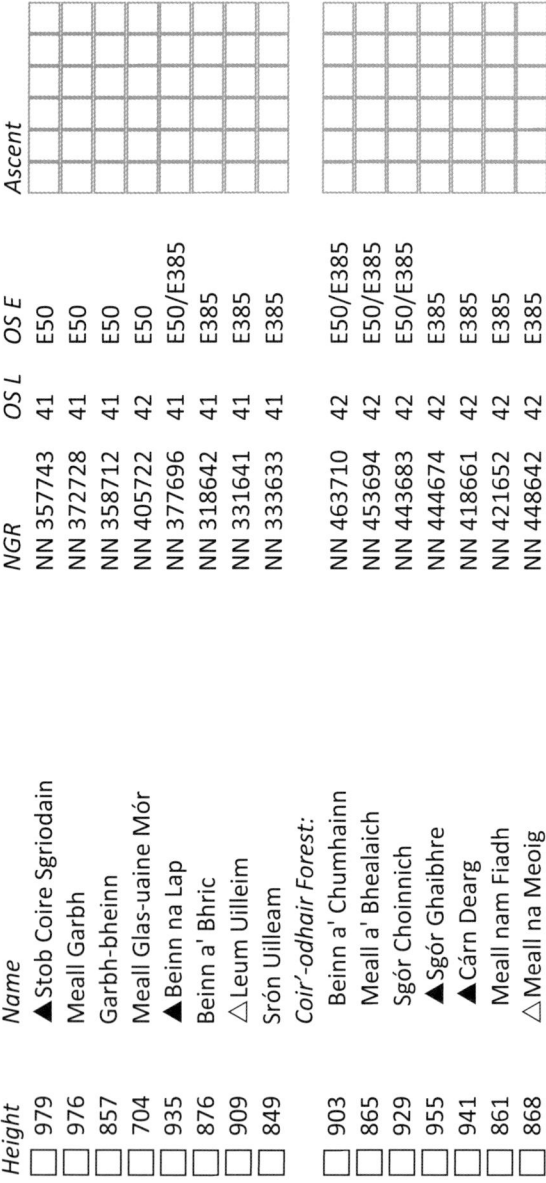

☐ 807   Beinn Pharlagain   NN 445632   42   E385

North Rannoch

This section sits astride the Atholl/Badenoch boundary, two mountain summits recording this in their names—the Boar of Badenoch and the Sow of Atholl. These peaks stand on their own at either side of Coire Dhomhain and are prominent from the Pass of Drumochter, the great road north. This part of the range is included within the Cairngorms National Park. On the Sow of Atholl is an area where Norwegian Blue Heather grows, one of very few patches in Scotland. The Sow also merits inclusion in Corbett's list.

Behind the two mountains rise four Munros which are possible to climb on a single day, Beinn Udlamain the highest at 3,306 feet. From Balsporran Cottages a path ascends the glen of the All Coire Fhár. At the head of the glen, when Loch Ericht comes into view, a short detour can be made northwards to take in Geal-charn. A' Mharconaich rises in a similar fashion on the south side of the glen, its summit reckoned to be at the north-east end of the flat ridge-top. The way is then made along the ridge to Beinn Udlamain, which is the only 1,000-metre peak in this section. The views over Loch Ericht to Ben Alder and its surroundings are tremendous. The walk continues down the south shoulder to Cárn 'Ic Loumhaidh. Sgairneach Mhór is quickly climbed from here, the smooth slopes easily ascended to the triangulation point on the top, almost on the edge of Coire Creagach. The way back is made by descending into Coire Dhomhain, though fitter walkers will want to include the Corbett of The Sow. The mountains are rather rounded, but here and there the rocks burst through the thin soil, particularly in the corries.

There are four Corbetts in this section, Stob an Aonaich Mhóir rising steeply from the side of Loch Ericht, though most folk will climb it from the east, where a track passes by, making its way from Bridge of Ericht to Coire Bhachdaidh Lodge. Beinn Mholach can be reach from Creaganour Lodge, a track heading into the hills for five miles from there. Beinn Mholach is a simple climb from Duinish at the head of Loch Garry over Creag nan Gabhar. There are a few other lower mountains in this area, including Beinn Bhoidheach and Glas Mheall Mór

16

Meall na Leitreach rises east of Loch Garry, a natural loch which is the source of the River Garry. It is fairly quickly climbed from Dalnaspidal Lodge.

The mountains decrease in altitude the nearer one gets towards Loch Rannoch, until they taper out into high moorland. A number of tracks and stalkers' paths give access from Rannoch-side, but they make long walks in. Hereabouts only Beinn a' Chuallaich (a Corbett) is of any consequence, a path from near Lochgarry House reaching the summit fairly steadily. This peak fails to be a Munro by only 73 feet. Loch Errochty to the north was extended by the erection of Errochty dam, 1,164 feet long and 162 feet high. Its waters are used in the generation of electricity at the power station near Tummel Bridge. At its west end rises a rounded summit, Creag a' Mhadaidh, which only just attains 2,000 feet. Another summit that just surpasses the 2,000 foot contour lies to the west, Gualann Sheileach.

Loch Garry, from where the River Garry rises, is a natural loch, attractive in its steep glacial valley. Dalnaspidal Lodge stands near its foot, by the side of the busy A9. A track from this lodge passes along the west shore of the loch. The western boundary is marked by Loch Ericht, a fifteen mile-long loch that was naturally formed but which was deepened by 26 feet when two dams were constructed at its opposite ends. Its surface level is 1,179 feet above sea level, and it is said that the waters never freeze. Some of the waters are piped down to the shores of Loch Rannoch, where a hydro-electric power station exists.

Almost all of the land within this section is owned by private estates, with the exception of some forestry land, which forms part of the Tay Forest Park. These blocks of forestry are located alongside Loch Rannoch.

16

Height	Name	NGR	OS L	OS E	Ascent
772	Creagan Mór	NN 615805	42	E50	
917	▲ Geal-charn	NN 598783	42	E50	
	Dalnaspidal Forest:				
739	An Torc (Boar of Badenoch)	NN 621762	42	E50	
975	▲A' Mharconaich (Bruach nan Iomairean)	NN 604763	42	E50	
1011	▲ Beinn Udlamain	NN 579740	42	E50	
991	▲ Sgairneach Mhór	NN 599731	42	E50	
803	△Meall an Dobharchain (The Sow of Atholl)	NN 624741	42	E50	
879	Meallan Buidhe	NN 611713	42	E50/E385	
	Talla Bheith Forest:				
865	An Sgulan	NN 569722	42	E50/E385	
827	Glas Mheall a' Chumhainn	NN 572696	42	E50/E385	
743	Stob Loch Monaidh	NN 559687	42	E50/E385	
855	△Stob an Aonaich Mhóir	NN 537693	42	E50/E385	
786	Cárn Dearg	NN 533675	42	E385	
625	Srón a' Chlaonaidh	NN 515654	42	E385	

	Height	Name	Grid ref		
☐	828	Glas Mheall Mór	NN 564673	42	E385
☐	789	Beinn Bhoidheach	NN 567657	42	E385
		Creagan Odhar Forest:			
☐	841	△Beinn Mholach	NN 587654	42	E385
☐	612	Gualann Sheileach	NN 616652	42	E49/E385
☐	612	Creag a' Mhadaidh	NN 634650	42	E385
☐	748	Càrn Fiaclach	NN 661621	42	E385
☐	892	△Beinn a' Chuallaich	NN 683618	42	E385
☐	743	Càrn Loch na Caillich	NN 697629	42	E385
☐	685	Meall Breac	NN 668695	42	E49/E51
☐	775	△Meall na Leitreach	NN 641702	42	E49/E51

Gaig & Atholl

Stretching over the border between Inverness and Perth shires are the great deer forests of Gaig and Atholl. These are wild, open places, high but generally flatter than many of the surrounding mountains. Little tree cover exists, other than plantations around the periphery.

The Gaig, or Gaick Forest, lies around Glen Tromie, which is an attractive glen with a track through it from the north. The lower hills soon give way to lesser peaks, some of which would be popular summits in other sections, but which hereabouts are lost in a sea of higher mountains. Little mountains such as Clach-mheall Dubh and Creag Ruadh are by-passed by walkers intent on ascending the lonely Munros in this section, such as the rounded top of Meall Chuaich. Paths or tracks from Cuaich can be followed to the little dam at Loch Cuaich, after which a steady pull is needed to the summit cairn.

The hills around Loch an t-Seilich, where Gaick Lodge is located, are steep-sided, making them more attractive-looking than their flatter plateau summits appear. Sgór Dearg is an attractive conical peak, a stalker's path zig-zagging up its steep eastern face.

To the south and west of here the high plateau continues, randomly rising above 3,000 feet to produce a Munro or two. The geography of the plateau means that the re-ascent of many summits is comparatively low, so that only two of the high tops meet Munro's standards, and having a low re-ascent, the lower summits do not qualify as Corbetts. The Munros are Cárn na Caim and A' Bhuidheanach Bheag. East of the latter is a Munro Top—Glas Mheall Mór. Glas Mheall Beag is a lower, but fairly attractive hill-top, and the two summits can be reached by walking a flattish ridge from Dalnaspidal on the A9.

To either side of Loch an Dúin the hillsides rise steeply, broken here and there with small outcrops and scree. An Dún rises to the west, a Corbett summit. The rocks of Creag an Loch rise on the east side of the loch and on the plateau, above the highest point, is A' Chaoirnich, a second Corbett.

The plateau spreading east is wild and open countryside, rising and falling between glens that sometimes cut deeply into the landscape, in other cases being shallow depressions, here and there marshy. Few of the summits are of any great significance. Leathad an Taobhain qualifies as a Corbett—in fact, an extra nine feet and it would be a Munro.

A series of glens from Blair Atholl and other locations alongside Glen Garry cut into the mountains from the south, giving different access routes to the tops. Popular is the route to Beinn Dearg, a solitary Munro that rises over 1,000 metres. Its narrow shape when looked from the south-west makes it conical in appearance. To its east is another elongated summit, Beinn Mheadhonach, a Corbett that rises over 900 metres. Northwards is Beinn Bhreac, another Corbett that almost reaches 3,000 feet but falls eight feet short.

Again lonely tops spread out to the north, rising eventually to over 1,000 metres at An Sgarsoch, and to its west another Munro, Cárn an Fhidhleir. The summit cairn marks the junction of Perth, Inverness and Aberdeen shires.

Glen Tilt forms the south-eastern boundary of this section. To its north rise steep slopes, at Cárn a' Chlamain attaining Munro height. The cairn on the summit is easily reached using a stalker's track from near Marble Lodge.

A pair of Corbett mountains rise to the east of Glen Tromie. Meallach Mhór and Cárn Dearg Mór are great rounded bulks that rise above their respective glens, Tromie to the west and Feshie to the east. Cárn Dearg Mór is the more attractive top, and higher of the two, which affords extensive views south and east to the loftier mountains of the Cairngorms National Park, which all of this section falls within.

Most of this section comprises sporting estates—the large Blair Castle property to the south, Gaick, Glen Feshie, and some others around the edge.

Height	Name	NGR	OS L	OS E	Ascent
	Gaig:				
619	Clach-mheall Dubh	NN 727904	35	E56	
644	Druim nan Sac	NN 703900	42	E51/E56	
658	Creag Ruadh	NN 684882	42	E51/E56	
951	▲ Meall Chuaich	NN 716879	42	E51	
898	Bogh-cloiche	NN 740867	42	E51	
801	Sgór Dearg	NN 747847	42	E51	
840	Creag an Dubh-chadha	NN 733829	42	E51	
882	A' Mharconaich	NN 708849	42	E51	
941	▲ Cárn na Caim	NN 677822	42	E51	
827	△ An Dun	NN 716802	42	E51	
875	△ A' Chaoirnich	NN 734802	42	E51	
853	Srón Bhuirich	NN 754824	42	E51	
782	Leathad na Lice	NN 762808	42	E51	
769	Bac na Creige	NN 774799	42	E51	
892	A' Chioch	NN 793843	42	E51	

17

☐	703	Meall an Dubh-chadha	NN 789908	35	E56
☐	769	△ Meallach Mhór	NN 776908	35	E56
☐	695	Meallach Bheag	NN 771921	35	E56
☐	626	Clach-mheall	NN 779931	35	E56
☐	643	Croidh-la	NN 776949	35	E56

Dalnacardoch Forest:

☐	915	Meall Coire nan Cisteachan	NN 664807	42	E51
☐	891 est	Meall Odhar Mór	NN 681803	42	E51
☐	936	▲ A' Bhuidheanach Bheag	NN 661777	42	E51
☐	928	Glas Mheall Mór	NN 681769	42	E51
☐	881	Glas Mheall Beag	NN 673757	42	E51
☐	803	Am Meadar	NN 702789	42	E51

Dail-na-mine Forest:

☐	621	Meall na Spianaig	NN 721776	42	E51
☐	793	Cárn a' Choire Odhair	NN 743786	42	E51
☐	816	Srón a' Chleirich	NN 784769	42	E51
☐	834	Meall Odhar a' Chire	NN 796787	42	E51

125

17

Height	Name	NGR	OS L	OS E	Ascent
791	Srón na Faiceachan	NN 808 775	43	E51	
782	Meall Odhar Ailleag	NN 809795	43	E51	
657	Cárn Dearg Mór	NN 851731	43	E51	
620	Elrig	NN 871725	43	E51	
899	Beinn a' Chait	NN 864748	43	E51	
1008	▲Beinn Dearg	NN 853778	43	E51	
786	Beinn Losgairnaich	NN 839788	43	E51	
885	Elrig 'ic an Toisich	NN 867788	43	E51	
870	Cárn a Chiaraidh	NN 879773	43	E51	
901	△Beinn Mheadhonach	NN 880758	43	E51	
	Coille Tarbh:				
879	Bráigh Srón Ghorm	NN 903783	43	E51	
887	Bráigh nan Creagan Breac	NN 898758	43	E51	
963	▲Cárn a' Chlamain	NN 916758	43	E51	
801 est	Meall Tionail	NN 920774	43	E51	
861 est	Conlach Mhór	NN 931768	43	E51	

	Height	Name	Grid Ref		
☐	729	Creag a' Chrochaidh	NN 946762	43	E51
☐	835	Bráigh Coire na Conlaich	NN 939778	43	E51
☐	692	Dun Mór	NN 966793	43	E51
☐	751 est	Stob Coire an Loch	NN 981835	43	E51
☐	799	Bráigh Coire Caochan nan Laoigh	NN 960818	43	E51
☐	1006	▲An Sgarsoch	NN 934837	43	E51
☐	774	Sgarsoch Bheag	NN 934852	43	E51
☐	705 est	Meall Dubh-chais	NN 921796	43	E51
☐	772	Cárn nan Airigh	NN 920807	43	E51
☐	994	▲Cárn an Fhidhleir (Cárn Ealar)	NN 904842	43	E51
☐	899	Meall Tionail	NN 891848	43	E51
☐	853	Srón Gharbh	NN 897822	43	E51
☐	692	Tom Liath	NN 891808	43	E51
☐	790	Meall Tionail na Beinne Brice	NN 882815	43	E51
☐	912	△Beinn Bhreac	NN 868821	43	E51
☐	837	Stob Coire nan Cisteachan	NN 864832	43	E51

17

Height	Name	NGR	OS L	OS E	Ascent
	Western Glen Feshie Forest:				
859	Glas-leathad Feshie	NN 845839	43	E51	
792	Uchd a' Chlársair	NN 816821	43	E51	
864	Stob Choire Bhran	NN 806946	43	E51	
912	△Leathad an Taobhain	NN 822858	43	E51	
831	Leac an Taobhain	NN 822849	43	E51	
847	Meall an Uillt Chreagaich	NN 826871	43	E51	
841	Glas-leathad Lorgaidh	NN 834871	43	E51	
849	Cárn an Fhidhleir Lorgaidh	NN 856875	43	E51/E57	
734	An Eilrig	NN 861860	43	E51	
733	Srón na Ban-righ	NN 877882	43	E51/E57	
712	Srón Direachain	NN 848902	35/43	E56/E57	
739	Srón na h-Iolaire	NN 832897	43	E51/E56	
857	△Cárn Dearg Mór	NN 823912	35/43	E56	
616	Aonach Mór	NN 807943	35/43	E56	
632	Meall Buidhe	NN 900955	35/43	E56	

Monadh Ruadh

The Monadh Ruadh is the traditional name for that part of the Grampian Highlands that stretch from Glen Feshie eastwards to the Cock Bridge to Tomintoul road. Many will refer to this area as the Cairngorm Mountains, but technically that name only applies to the Cárn Gorm to Beinn MacDuibh massif. Today, however, the term Cairngorms is applied to a much wider area, the Cairngorms National Park including the surrounding ranges.

Many of Britain's highest mountains are located in this area, with the second highest summit in Britain—Beinn MacDuibh—being here, with a number of other summits that are in excess of 1,000 metres. Of these, Cárn Gorm, or Cairn Gorm itself is the most popular climbed, though many will use the skiers' access road and funicular railway to travel to the Ptarmigan Station before walking the final stretch to the summit. On clear days the views are extensive.

From Cárn Gorm an elevated plateau with various summits can be traversed, its sub-arctic geology, fauna and weather being distinctive in this country. Stob Coire an t-Sneachda and Cárn Lochain form attractive summits. Beinn MacDuibh to the south of here is often approached from this side, or else from Braemar and the Lairig Ghru.

Around it are a series of subsidiary mountains, each interesting in their own right, and in some cases meriting inclusion in Munro's list. An example is Cárn a' Mhaim, which elsewhere in Scotland would be a notable peak, but which here is subsidiary to higher mountain tops.

West of the Lairig Ghru are the summits of Bráigh Riabhach, Cárn an t-Sabhail and Sgór an Lochain Uaine, significant and attractive mountains with steep sides, particularly the eastern corries. Bod an Deamhain is a small, but startling high summit above the Lairig Ghru.

In the basin of Garbh Choire Mór, to the east

of Cárn na Criche, a snow patch often survives from one winter to the next. This is one of the few places in Britain where this occurs, and the patch of snow has been named the Sphinx. It is regarded as being Britain's most durable snow patch, only fully melting eight times since 1850.

Beyond the Bráigh Riabhach massif are a number of other 1,000 metres summits, including Beinn Bhrotain, Cárn Bán Mór, Sgór Gaoith and Mullach Clach a' Bhláir. The summits west of Loch Eanaich are the more interesting, those further south being little more than excessively high plateaus.

East of the Cairngorm Mountains is a wild stretch of mountainous countryside, with a number of major summits in it. North of Braemar rises Beinn a' Bhuird and Ben Avon, two significant mountains, in excess of 1,100 metres. Beinn a' Bhuird is noted for its deep rock-girt corries, Ben Avon for its rocky tors on the summit ridge. Adjoining them are many lesser summits, a few of them of Corbett status. Around Glen Avon are many tops in excess of 2,000 feet, the extensive plateau here riven with deep glens which are probably more attractive than the flatter summits. There are some exceptions, such as the conical Culardoch, which almost reaches Munro height.

North of Glen More and the Queen's Forest rise the Kincardine Hills, the highest of which—Meall a' Bhuachaille—is a Corbett. These form an easier and lower hill-walk than the higher summits to the south.

Most of the Forest of Mar and the countryside west of Beinn a' Bhuird forms part of the National Trust for Scotland's Mar Lodge estate. Cairn Gorm and the northern corries are owned by Highlands and Islands Enterprise. Another area in public ownership is upper Glenavon, around Loch Avon, which is the property of the Royal Society for the Protection of Birds. Rothiemurchus estate to the west is private, as are the estates to the east, such as Glenavon, Invercauld and Delnadamph, the latter part of the Royal Family's Balmoral estate.

18

Height	Name	NGR	OS L	OS E	Ascent
	Kincardine Hills:				
711	Creag a' Chaillich	NH 968127	36	E57	
732	Creagan Gorm	NH 978120	36	E57	
810	△ Meall a' Bhuachaille	NH 991116	36	E57	
848	Creag Dhubh	NH 906043	36	E57	
1111	Sgóran Dubh Mór	NH 904002	36	E57	
976	Meall Buidhe	NH 892001	36	E57	
920	Geal-charn	NH 884014	36	E57	
742	Creag Mhigeachaidh	NH 873023	36	E57	
1118	▲ Sgór Gaoith	NN 902989	36	E57	
1052	Cárn Ban Mór	NN 893971	35/36/43	E57	
998	Meall Dubhag	NN 881956	35/36/43	E57	
800	Meall nan Sleac	NN 868944	35/36/43	E57	
1019	▲ Mullach Clach a' Bhlair	NN 883927	35/43	E57	
742	Cárn Eilrig	NH 938053	36	E57	
734	Cárn Odhar	NH 947041	36	E57	

18

Height	Name	NGR	OS L	OS E	Ascent
1184	Srón na Lairige	NH 964006	36	E57	
1296	▲ Bràigh Riabhach (Braeriach)	NN 954999	36/43	E57	
1265	Càrn na Criche	NN 939982	36/43	E57	
1258	▲ Sgòr an Lochan Uaine	NN 954977	36/43	E57	
1291	▲ Càrn an t-Sabhail (Cairn Toul)	NN 963973	36/43	E57	
1213	Stob Coire an t-Saighdeir	NN 962963	36/43	E57	
1004	▲ Bod an Deamhain (Devil's Point)	NN 976951	36/43	E57	
918	Tom Dubh	NN 921952	36/43	E57	
1113	▲ Monadh Mór	NN 938942	36/43	E57	
1157	▲ Beinn Bhrotain	NN 954923	43	E57	
942	Càrn Cloich-mhuilinn	NN 968907	43	E57	
623	Cairn Geldie (Cnapan Or)	NN 995885	43	E51/E57	
622	Creag nan Gall	NJ 005102	36	E57	
742	Stac na h-Iolaire	NJ 017089	36	E57	
	Cairngorm Mountains:				
1244	▲ Càrn Gorm (Cairn Gorm)	NJ 005041	36	E57	

18

	Height	Name	Grid Ref		
☐	1176	Stob Coire an t-Sneachda	NH 996030	36	E57
☐	1215	Cárn Lochain	NH 986026	36	E57
☐	1053	Creag an Leth-choin	NH 969033	36	E57
☐	787	Creag a' Chalamain	NH 962054	36	E57
☐	644	Airgiod-meall	NH 966067	36	E57
☐	1309	▲Beinn MacDuibh (Ben Macdui)	NN 989989	36/43	E57
☐	1037	▲Cárn a' Mhaim	NN 994952	36/43	E57
☐	1108	Creagan a' Choire Etchachan	NO 011996	36/43	E57
☐	1182	▲Beinn Mheadhoin	NJ 024017	36	E57
☐	983	Sgurr an Lochan Uaine	NO 025990	36/43	E57
☐	1155	▲Doire Cárn Gorm (Derry Cairngorm)	NO 016981	36/43	E57
☐	890	Cárn Crom	NO 023954	36/43	E57
☐	722	Creagan nan Gabhar	NN 999923	43	E57
☐	813	△Sgór Mór	NO 007914	43	E53/E57
☐	741	Sgór Dubh	NO 034921	43	E57/E58
☐	1016	A' Choinneach	NJ 032048	36	E57/E58
☐	1090	▲Beinn Beithneag (Bynack More)	NJ 042063	36	E57/E58

The Mountains of Great Britain

Height	Name	NGR	OS L	OS E	Ascent
895	△ Creag Mhór	NJ 057048	36	E57/E58	
736	Monadh nan Eun	NJ 108082	36	E58	
742	Big Garvoun	NJ 148084	36	E58	
715	Cnap Chaochan Aitinn	NJ 146100	36	E58	
691	Cárn na Ruabraich	NJ 129108	36	E58	
616	Cárn Ruabraich	NJ 136124	36	E58	
804	Cárn Bheadhair	NJ 055117	36	E57/E58	
821	△ Geal Charn	NJ 090127	36	E58	
637	Cárn na h-Ailig	NJ 073140	36	E58	
688	Cárn na Farraidh	NJ 114148	36	E58	
1083	▲ Beinn a' Chaorainn	NJ 045014	36	E57/E58	
1017	Beinn a' Chaorainn Bheag	NJ 057017	36	E57/E58	
931	▲ Beinn Bhreac	NO 059971	36/43	E57/E58	
777	Meall an Lundain	NO 062948	36/43	E57/E58	
668	Creag Bhalg	NO 092912	43	E52/E58	
1179	Beinn a' Bhuird Deas	NO 090979	36/43	E58	

☐	1197	▲ Beinn a' Bhuird	NJ 092006	36	E58
☐	715	The Bruach	NJ 118056	36	E58

Ben Avon:

☐	972	Creag an Dail Mhór	NO 132982	36/43	E58
☐	1171	▲ Ben Avon (Leabaidh an Daimh Bhuidhe)	NJ 132018	36	E58
☐	1023	Bráigh Mór (West Meur Gorm Craig)	NJ 154036	36	E58
☐	912	Meall Gaineimh	NJ 167051	36	E58
☐	629	Meall an t-Seangain	NJ 179050	36	E58
☐	772	Cárn Tiekeiver	NJ 176022	36	E58
☐	650	Creag a' Chleirich	NO 141933	36/43	E58
☐	818	△ Cárn na Drochaide	NO 127938	36/43	E58
☐	733	Meall an t-Slugain	NO 126958	36/43	E58
☐	863	△ Creag an Dail Bheag	NO 157982	36/43	E58
☐	861	Cárn Liath	NO 164977	36/43	E58
☐	649	Creag a' Chait	NO 172959	36/43	E53/E58
☐	617	Meall Gorm	NO 185945	36/43	E53/E58
☐	635	Creag Leac (Craig Leak)	NO 185931	36/43	E53/E58

135

18

Height	Name	NGR	OS L	OS E
900	△ Cul Ardach Mór (Culardoch)	NO 193988	36/43	E58
696	Tom Breac	NO 222998	36/43	E58
657	Cárn a' Mhadaidh (Fox Cairn)	NJ 286028	37	E59
	Dail-na-damh Forest:			
745	Cárn a' Bhacain	NJ 291043	37	E59
694	Camock Hill	NJ 273146	37	E59
680	An Ca	NJ 281052	37	E59
699	Cárn Leac Saighdeir	NJ 271069	37	E59
704	Cárn Oighreag	NJ 240071	36	E58
829	△ Cárn na Saobhaidh (Cairn Sawvie)	NJ 221044	36	E58
802	Geal Charn Mór	NJ 200052	36	E58
726	Cárn Culchabhie	NJ 200071	36	E58
650	Cárn Loch an Builg	NJ 201035	36	E58
629	Cárn Bad a' Ghuail	NJ 179008	36	E58
711	Creag Mheann (Craig Veann)	NJ 188110	36	E58
693	Tolm Buirich	NJ 211121	36	E58

Ascent

18

Height	Name	NGR	OS L	OS E	Ascent
☐ 792	Cárn Ealasaid	NJ 228118	36	E58	
☐ 776	Beinn a' Chruinnich	NJ 237132	36	E58/E62	
☐ 664	Liath Bheinn	NJ 172122	36	E58	

Glengairn, Ladder & Cromdale

This section includes those mountains that are located in the Grampian hills to the north and east of the A939 road that links Ballater with Grantown-on-Spey. In many cases a continuation of the Monadh Ruadh, these summits are generally lower and in some cases have their own group names, such as the Hills of Cromdale and the Ladder Hills.

Beinn Rinneis, or Ben Rinnes, is one of the higher summits, standing alone between Dufftown and Glenlivet. All of this is Speyside whisky country, and Ben Rinnes has a distillery of that name at its foot. Being so distinctive in the surrounding area, this is a popular climb, the summit of the Corbett being located at the Scurran of Lochterlandoch. Various paths and tracks lead to the top from different points below.

Three of the Hills of Cromdale meet the specification for inclusion within these tables. Cárn a' Ghille Chearr is located at the northern end of the range, and the summit affords extensive views over Moray. The summit forms the northern boundary of the Cairngorms National Park. The highest point in the range is Creagan a' Chaise, where the Jubilee Cairn marks the top. This hill is easily climbed from Lynemore to the west, from where tracks lead across the moors and up towards the top, giving access to various grouse butts. It crosses Cárn Tuairneir, the third Cromdale summit.

Around Glen Fiddich and the Blackwater Forest are a series of summits in excess of 2,000 feet. Corryhabbie Hill is a Corbett, the track known as Morton's Way making a route to the top. Cárn a' Bhodaich is a top that merits inclusion due to its re-ascent, whereas other higher summits on the ridge south of Corryhabbie are not prominent enough to count.

In the Blackwater Forest rise Cook's Cairn, Cárn na Bruar, and a few lesser summits. The glen east and north of Cook's Cairn has been covered by wind turbines, industrialising the landscape.

The hills to the south-west of Round Hill are known as the Ladder Hills, falling within the Cairngorms National Park. Cárn Mór is the tallest, a Corbett, rising steeply above the Braes of Glenlivet from where a short but steep ascent can be made from Scalan, near Chapeltown of Glenlivet. Most walkers will prefer to make a traverse of the ridge, claiming the various tops. At the west end of the range is Coir' Riabhach Mór, with its ski tow on the western slopes, part of the Lecht Ski Centre. Geal Charn rises above Corgarff, a shooter's track leading to near the summit from the south.

In the Cabrach to Glen Buchat stretch of hills are a few summits that exceed 2,000 feet. Again, much of this area is protected for grouse shooting, and out of season the various tracks to the grouse butts allow easy walking. Creag an Sgór and Creag an Eunan are two summits that are quickly climbed by track from Glen Buchat. At the Cabrach end of the range rises The Buck, a prominent summit hereabouts. Being the highest point in the immediate locality means that it affords extensive views across the Aberdeenshire hills, the rich farming and forestry landscape of Strathbogie and Garioch spread out below.

The Glengairn Hills are located to either side of that glen. Geallaig Hill is a prominent summit to the north of the River Dee, a track passing over the top which allows fairly easy walking. North-east of Glen Gairn itself is a range of lesser tops, extending from Cairnagour Hill in the west to Mullachdubh in the east. Within the range is Monadh Gobhann (or Mona Gowan) the summit marked by a Jubilee Cairn.

South-east of Mullachdubh rises Mór Bheinn (Morven), a prominent Corbett, the summit of which is marked by a prehistoric cairn. Farther east, on the northern periphery of the Howe of Cromar, rises Pressendye, the summit cairn there protruding above 2,000 feet. Most of the land within this section forms extensive sporting estates, some of it being Crown Estate property.

19

Height	Name	NGR	OS L	OS E	Ascent
☐ 841	△ Beinn Rinneis (Ben Rinnes)	NJ 255354	28	E424	
	Hills of Cromdale:				
☐ 710	Cárn a' Ghille Chearr	NJ 139298	36	E61	
☐ 722	Creagan a' Chaise	NJ 104242	36	E61	
☐ 693	Cárn Tuairneir	NJ 098233	36	E61	
☐ 781	△ Corryhabbie Hill	NJ 281289	37	E62	
☐ 655	Cárn a' Bhodaich	NJ 260289	37	E62	
	Blackwater Forest:				
☐ 755	Cook's Cairn	NJ 300275	37	E62	
☐ 683	Cárn na Bruar	NJ 291255	37	E62	
☐ 668	Round Hill	NJ 306226	37	E62	
☐ 632	Hill of Three Stones	NJ 349225	37	E62	
☐ 619	Cairnbrallan	NJ 333245	37	E62	
	Ladder Hills:				
☐ 742	Geal Charn Beag	NJ 298197	37	E62	
☐ 788	Letterach (Geal Charn Mór)	NJ 283206	37	E62	

	Name	Grid Ref		
659	Moss Hill	NJ 316173	37	E62
617	Finlate Hill	NJ 293179	37	E62
804	△ Cárn Mór	NJ 266183	37	E62
800	Monadh an t-Sluichd Leith	NJ 262171	37	E62
792	Cárn Liath	NJ 253159	37	E58/E62
718	An Socach	NJ 277143	37	E59/E62
636	Cnoc nan Dubh Breac	NJ 297152	37	E59/E62
779	Coir' Riabhach Mór	NJ 252134	37	E58/E62
674	Geal Charn	NJ 284109	37	E59
634	Creag an Sgór	NJ 375196	37	E62
633	Creag an Eunan	NJ 386191	37	E62
721	The Buck	NJ 413234	37	E62
	Glengairn Hills:			
743	Geallaig Hill	NO 298982	37	E62
743	Cairnagour Hill	NJ 325056	37	E62
638	Tom Liath	NJ 332041	37	E62
749	Monadh Gobhann	NJ 336058	37	E62

19

19

Height			Name	NGR	OS L	OS E	Ascent
☐ 681			Mullachdubh	NJ 354067	37	E62	
☐ 872			△ Mór Bheinn (Morven)	NJ 377040	37	E62	
☐ 619			Pressendye	NJ 490090	37	E62	

Fearnach & Ey

From Pitlochry to Braemar is a high section of the Grampian Mountains, spreading east from Glen Tilt to Glen Shee. Within the area are a number of high Munros, though few of them are well known, other than the ski mountains above Glen Shee and Beinn a' Ghlo to the south.

Cárn Aosda and The Cairnwell both rise over 3,000 feet, the former by as little as three feet. The southern slopes of Cárn Aosda and the northern slopes of The Cairnwell have a series of ski tows and chair lifts on them making the area very popular for winter sports. Tracks from the ski centre, established in 1962, make ascents of the hills fairly straightforward. Similarly, a track can be followed from Glen Clunie to the top of Mór Shrón, or Morrone, above Braemar, though the path up the Braemar face is preferable and is used for the annual hill-race. This summit is a Corbett and an excellent viewpoint for views of the wider Cairngorms National Park, within which most of this section lies.

The range of hills spreading south-west from Morrone gradually increases in height over Sgór Mór to An Socach, a twin-topped Munro, sometimes referred to as Socach Mór. The western top is the Munro, a scree-spattered hill. On the other side of Glen Ey from here are a few more Munro summits located within a sea of lesser mountains. The random form of the hillsides here makes single linear walks to pick off tops rather difficult, the climber requiring to nip out to summits on side ridges, such as Buachaille Bréige and Meall Tionail.

Cárn Bhac is the first Munro in this part of the section, again a fairly stony summit, rounded on the top with steep sides into the glens or corries to the side. Access to the top requires a long walk, often made from Inverey to the north.

The two Munros to the south of here, Beinn Iutharn Mhór and Cárn an Righ, are far more interesting summits, with steep corries and scree slopes on many of the sides. Mám nan Cárn and Beinn Iutharn Bheag are two other mountains listed as tops by Munro, and these are popular with Munro-baggers.

The greatest height in this part of the mountains is Glas Tulaichean, rising above 1,050 metres. Various tracks from Gleann Lochsaidh make their way to the summit which hangs over Glas Choire Mhór. This corrie faces into Gleann Taitneach, on the other side of which rises Cárn Bhinnein (a Munro top) Cárn a' Gheóidh (a Munro) and a second top, Cárn nan Sac. This brings us back round to Glen Shee.

A range of lonely hills spreads south-eastwards on the south side of Gleann Lochsaidh. Although some rise over 2,500 feet, such as Meall a' Choire Bhuidhe and Beinn Earb, they fail to have sufficient re-ascent to merit listing as Corbetts. On the north side of the glen, rising steeply above the Spittal of Glenshee, is Beinn Gulabin, which does meet Corbett's criteria.

Rising to the north-east of Blair Atholl is the great mountain of Beinn a' Ghló. This is really a series of mountains, starting with Cárn Liath to the south-west, which merits listing as a Munro. A well-used path leads on to Bráigh Coire Chruinn-bhalgain, a second Munro and rising to 3,505 feet. A steep descent to Bealach an Fhiodha and a climb up to Cárn nan Gabhar adds a third Munro, and the tallest part of Beinn a' Ghló. Airgiod Bheinn on the same ridge is a Munro top. This mountain has been recognised for its rare Alpine plants.

From the latter mountain one can look south and east towards Beinn Bhuirich, or Ben Vurich, a conical Corbett that fails to reach Munro status by 38 feet, as well as Beinn Bhreacaidh, or Ben Vrackie, a second Corbett that is much associated with Pitlochry, rising as it does to the north of that town, and from where a footpath makes its way to the top.

Most of this mountain country is owned by private sporting estates, much of it being part of Atholl Estates to the west, Invercauld Estates to the east, and various other owners across the centre.

Height	Name	NGR	OS L	OS E	Ascent
859	△Mór Shron (Mór Bheinn, Morrone or Morven)	NO 132887	43	E52/E58	
830	Cárn na Drochaide	NO 125861	43	E52	
706	Cárn Mór	NO 102871	43	E52	
635	Tom Anthon	NO 098881	43	E52/E58	
792	Sgór Beag	NO 111842	43	E52	
887	Sgór Mór	NO 116826	43	E52	
938	An Socach Ear	NO 099806	43	E52	
944	▲An Socach (Socach Mór)	NO 080800	43	E52	
894	Cárn Creagach	NO 069830	43	E52	
946	▲Cárn Bhac	NO 051832	43	E52	
878	Geal Charn	NO 032833	43	E52	
712	Meall Tionail	NO 012847	43	E52	
745	Buachaille Breige	NO 022854	43	E52	
719	Cárn an t-Sionnaich	NO 017860	43	E52	
797	Meall Christie	NN 044860	43	E52	
818	Cárn Liath	NO 036867	43	E52	

20

Ascent

Height	Name	NGR	OS L	OS E
784	Cárn Damhaireach (Top of the Battery)	NO 059859	43	E52
681	Meall Chrombaig	NO 008807	43	E52
1029	▲ Cárn an Righ	NO 029772	43	E52
986	Mám nan Cárn	NO 049779	43	E52
1045	▲ Beinn Iutharn Mhór	NO 046792	43	E52
953	Beinn Iutharn Bheag	NO 065791	43	E52
915	▲ Cárn Aosda	NO 134792	43	E52
933	▲ The Cairnwell (An Cárn Bhalg)	NO 134773	43	E52
975	▲ Cárn a' Gheoidh	NO 107767	43	E52
876	Cárn Mór	NO 110751	43	E52
806	△ Beinn Gulabin	NO 101722	43	E52
917	Cárn Bhinnein	NO 091763	43	E52
841	Creag Easgaidh	NO 077769	43	E52
871	Cárn a' Chlarsaich	NO 069779	43	E52
	Gleann Lochsaidh:			
858	Cárn a' Glas Choire Bheag	NO 061776	43	E52

20

1051	▲Glas Tulaichean	NO 051760	43	E52
827	Creag Bhreac Ard	NO 071742	43	E52
858	Cárn Dallaig	NO 016749	43	E52
770	Creag Dhubh Ard	NO 037731	43	E52
786	Cárn Dearg	NO 043719	43	E52
815	Meall Ruigh Mór Thearlaich	NO 052719	43	E52
868	Meall a' Choire Bhuidhe	NO 062710	43	E52
617	Beinn a' Chruachain	NO 047695	43	E52
802	Beinn Earb	NO 079691	43	E52
728	Creag an Dubh Shluic	NO 090688	43	E52

Glen Shee:

794	Meall Uaine	NO 111674	43	E52
686	Meall Odhar	NO 118658	43	E52

Gleann Loch:

763	Meall na Spionaig	NO 001775	43	E52
668	Cnapan Dubh	NO 003757	43	E52
647	Meall Reamhar	NN 994760	43	E51

20

20

Height	Name	NGR	OS L	OS E	Ascent
☐ 718	Creag Cam a' Choire	NO 003749	43	E52	
☐ 726	Creag Leacagach	NO 002729	43	E52	
☐ 759	Meall Gharran	NN 976762	43	E51	
☐ 1121	▲Beinn a' Ghlo (Cárn nan Gabhar)	NN 971733	43	E51	
☐ 1061	Airgiod Bheinn	NN 962720	43	E51	
☐ 672	Stac nam Bodach	NN 979702	43	E51	
☐ 903	△Beinn Bhuirich (Ben Vuirich)	NN 997700	43	E51	
☐ 635	Creag an t-Sithein	NO 032658	43	E52	
☐ 618	Crungie Clach	NN 987657	43	E49	
☐ 675	Meall Breac	NN 968688	43	E49	
	Killiecrankie Hills:				
☐ 776	Cárn Geal	NN 968633	43	E49	
☐ 841	△Beinn Bhreacaidh (Ben Vrackie)	NN 951632	43	E49	
☐ 722	Meall an Daimh	NN 939641	43	E49	
☐ 633	Meall na h-Aodainn Móire	NN 942622	43	E49	
☐ 627	Meall Uaine	NN 934616	43	E49	

20

641	☐	Blath Bhalg (Creag Dubh)	NO 019611	43	E52
1070	☐	▲ Bráigh Coire Chruinn-bhalgain	NN 945724	43	E51
737	☐	Beinn Bheag	NN 949704	43	E51
976	☐	▲ Cárn Liath	NN 936698	43	E51

Monadh Geal

The Monadh Geal is better known as the Mounth, an Anglicised version of its proper Gaelic name. Many of the hills have been renamed here, particularly to the east, the original Gaelic names, used not so long ago, forgotten, and rather comical-sounding English versions used in their place. Examples not included in the following list of mountains (where others can be found) include Hill of Badymicks, Glittering Skellies, Baudy Meg, Bonnyfleeces, Shank of Donald Young, and Auld Darkney. Farther west the original Gaelic names have survived.

Queen Victoria was a great lover of these mountains; she and Prince Albert purchased Balmoral Castle on the northern periphery in 1853 (having previously leased it) and rebuilt it in a neo-baronial Scots style. The present Queen still owns the castle, as well as 29,000 acres of mountainous country including Lochnagar and Loch Muick, and takes a holiday there every year. Queen Victoria climbed Lochnagar in 1848 and various other members of royalty have done so since.

Lochnagar (Beinn nan Ciochan) is the finest summit in the Mounth, its great northern corrie dropping in granite cliffs to Lochnagar itself, the lochan from which the mountain receives its name. Lord Byron wrote the famous lines:

England! Thy beauties are tame and domestic,
To one who has roved o'er the mountains afar:
Oh for the crags that are wild and majestic!
The steep frowning glories of dark Loch na Garr.

From the Spittal of Glenmuick a track then a path can be followed all the way to the top, passing a memorial on the slopes below Cuidhe Crom. The actual summit is at Cac Cárn Beag. A longer route can be made from Balmoral. The ring round Loch Muick can be continued from here, west to Cárn an t-Sagairt Mór then south over Cárn Bannoch and Cárn Braghad, from where a descent can be made to Loch Muick.

The northern cliff of Creag an Dubh-loch is a notable resort for climbers. Much of the high country hereabouts forms the Lochnagar Special Protection Area.

South-west of Lochnagar is the Caenlochan Special Area of Conservation, protecting the high summits and corries, home of Arctic plant life. Within the reserve is Glas Maol, easily reached from the west, there being plenty of parking at the Glenshee ski centre. The lower slopes of Glas Maol and Meall Odhar have ski tows on them. Glas Maol is a rounded hill, despite its height, whereas the lower Creag Leacach to the south-west is more prominent, and covered with scree. The deep Caenlochan Glen can be seen to the east of Glas Maol, with Canness Glen forming a huge corrie at the head of Glen Isla. Corrie Fee is a National Nature Reserve, being one of the finest corries in Britain, with notable montane willow scrub and the rare alpine coltsfoot.

Many hill-tracks and paths cross the Mounth, most of them having Mounth in their names. From east to west they are Cairn a Mounth, Fungle, Fir Mounth, Mounth Keen, Capel Mounth, Tolmounth and Monega. All are rights of way and popular with walkers. Mount Keen, an outlying Munro, is quickly ascended by means of the Mounth Keen path which passes below its western slopes. The path leaves the public road at Glenesk to the south, passing the Queen's Well in Glen Mark, where Victoria had a drink on one of her rambles. Loch Lee nearby is attractive in its steep-sided glen, surrounded by a number of summits that can readily be climbed.

To the south, Glen Clova is the most popularly used route to access these hills. The Munros of Mayar and Driesh are easily climbed from the glen, which has many rocky corries higher up its steep slopes. Paths also head towards Broad Cairn and Tolmount. Glas Maol and Cárn na Claise are reached from the ski centre at the head of Glen Clunie, or alternatively from upper Glen Isla, a quieter and more attractive approach.

21

Height	Name	NGR	OS L	OS E	Ascent
617	Peter Hill	NO 578886	44	E54	
778	△ Mount Battock	NO 549844	44	E54	
618	Hill of Cammie	NO 526854	44	E54	
688	Mudlee Bracks	NO 511857	44	E54	
723	Tampie	NO 497868	44	E54	
731	Gannoch	NO 497880	44	E54	
742	Hill of Cat (Cnoc an Cat)	NO 484872	44	E54	
727	Cock Cairn	NO 462887	44	E54	
756	Hill of Gairney	NO 448876	44	E54	
664	Cárn a' Ghlais-choire	NO 434842	44	E54	
887	Braid Cairn	NO 426872	44	E54	
939	▲ Mount Keen (Monadh Caoin)	NO 409869	44	E54	
626	Clachan Yell	NO 446911	44	E53	
640	Black Craig (Creag Dubh)	NO 431905	44	E53	
700	Cairn Leuchan	NO 379906	44	E53	
709	Creag Dearg	NO 360876	44	E53	

21

721	Fasheilach	NO 342857	44	E53
684	Cárn a' Choire Bhreac	NO 367847	44	E53
774	Black Hill of Mark	NO 324814	44	E53
834	Easter Balloch	NO 348802	44	E53
715	Wolf Craig (Creag a' Mhadaidh)	NO 380824	44	E53
696	Monawee	NO 409809	44	E54
705	Hunt Hill	NO 380805	44	E53
678	Hill of Wirren	NO 523739	44	E389
691	West Knock (Cnoc Iar)	NO 474757	44	E54/E389
695	Black Hill (Cnoc Dubh)	NO 458752	44	E54/E389
687	Cairn of Meadows	NO 435749	44	E389
741	Cruys	NO 421756	44	E54/E389
648	Cairn Caidloch	NO 431783	44	E54
663	Burnt Hill	NO 418779	44	E54
687	Craig Maskeldie	NO 391796	44	E53
682	Cairn Lick	NO 392782	44	E53
639	Wester Hill of Berran	NO 445713	44	E389

21

Ascent

Height	Name	NGR	OS L	OS E	Ascent
692	Earn Skelly	NO 436713	44	E389	
726	Hill of Glansie	NO 430698	44	E389	
735	Ruragh	NO 421711	44	E389	
756	Finbracks	NO 402703	44	E389	
685	Manywee	NO 392692	44	E53	
778	White Hill (Cnoc Ban)	NO 401731	44	E389	
896	Beinn Tirran (The Goet)	NO 374746	44	E53	
806	Wester Balloch	NO 341790	44	E53	
763	Wester Watery Knowe	NO 350782	44	E53	
870	Green Hill (Cnoc Uaine)	NO 348757	44	E53	
837	The Snub (An t-Srón)	NO 335757	44	E53	
841	Benty Roads	NO 331766	44	E53	
876	Boustie Ley	NO 322760	44	E53	
862	Lair of Whitestone (Cathelle Houses)	NO 312764	44	E53	
832	Lair of Aldarie	NO 312780	44	E53	
801	Ferrowie	NO 303794	44	E53	

21

	Height	Name	Grid Reference		Map
☐	754	Black Hill (Cnoc Dubh)	NO 311824	44	E53
☐	732	Dog Hillock (Cnocan an Cu)	NO 286794	44	E53
☐	768	Sandy Hillock	NO 266804	44	E53
		Balmoral Forest:			
☐	862	Caisteal na Caillich	NO 282875	44	E53
☐	865	△ Connachcreag Mór	NO 280865	44	E53
☐	980	Meikle Pap (Cioch Mhór)	NO 260861	44	E53
☐	1083	Cuidhe Crom	NO 260849	44	E53
☐	1156	▲ Lochnagar (Beinn nan Ciochan)	NO 244861	44	E53
☐	974	Meall Coire na Saobhaidhe	NO 243873	44	E53
☐	885	Meall an Tionail	NO 223877	44	E53
☐	830	Cnapan Nathraichean Mór	NO 229885	44	E53/E58
☐	824	Cnapan Nathraichean	NO 229885	44	E53/E58
☐	1110	▲ Cárn a' Choire Bhoidheach	NO 227846	44	E53
☐	1044	Cárn an t-Sagairt Beag	NO 216848	44	E53
☐	1047	▲ Cárn an t-Sagairt Mór	NO 208843	44	E53
☐	710	An t-Srón	NO 284844	44	E53

21

21

Height	Name	NGR	OS L	OS E	Ascent
836	Creag an Loch	NO 192848	43	E52/E53	
849	Meall an t-Slugain	NO 184863	43	E52/E53	
795	Creag Loisgte	NO 177868	43	E52/E53	
784	Creag na Leachda	NO 179887	43	E52/E53/E58	
743	Millstone Cairn	NO 170890	43	E52/E58	
690	Cárn nan Sgliat	NO 167902	43	E52/E58	
1012	▲Cárn Beannach (Cairn Bannoch)	NO 222826	44	E53	
983	Creag an Dubh-loch	NO 233823	44	E53	
998	▲Broad Cairn (Can Braghaid)	NO 240817	44	E53	
866	Creag Mellon	NO 262774	44	E53	
846	Cárn Damh (Cairn Damff)	NO 249777	44	E53	
874	Meall an Loch Esk	NO 229790	44	E53	
920	Crow Craigies	NO 222799	44	E53	
958	▲Tolmouint	NO 210800	44	E53	
957	▲Tom Buidhe	NO 213788	44	E53	
905	Finalty Hill	NO 212751	44	E53	

	Height	Name	Grid Ref		Map
☐	928	▲ Mayar	NO 241738	44	E53
☐	830	Bawhelps	NO 227722	44	E53
☐	689	Craigie Thieves	NO 243700	44	E53
☐	740	Badandun Hill	NO 207679	44	E53
☐	611	Corwharn	NO 288651	44	E53
☐	947	▲ Driesh (Dris)	NO 271735	44	E53
☐	850	Hill of Strone (Cárn na Srón)	NO 288729	44	E53
☐	651	Cairn of Barns	NO 320712	44	E53
☐	671	Cat Law (Cnoc nam Cat)	NO 319611	44	E53
☐	834	△ Creag nan Gabhar	NO 155841	43	E52
☐	822	Cárn Dubh	NO 162819	43	E52
☐	1019	▲ Cárn an Tuirc	NO 174804	43	E52/E53
☐	1064	▲ Cárn na Claise	NO 186789	43	E52/E53
☐	1068	▲ Glas Maol	NO 167766	43	E52
☐	987	▲ Creag Leacach	NO 155745	43	E52
☐	759	Meall Gorm	NO 140748	43	E52
☐	755	Cárn an Daimh	NO 136712	43	E52

21

21

Height	Name	NGR	OS L	OS E	Ascent
☐ 767	Mallrenheskein	NO 153727	43	E52	
☐ 757	Cnoc Dubh (Black Hill)	NO 163719	43	E52	
☐ 807	△Monadh Meadhonach (Monamenach)	NO 176707	43	E52/E53	
☐ 664	Creagan Caise	NO 181690	43	E52/E53	
☐ 702	Meall na Leitir (Duchray Hill)	NO 161762	43	E52	
☐ 744	Mount Blair	NO 168630	43	E52	

Skye

Some of the most distinct mountains in Britain are to be found on the island of Skye. The Cuillin rise from sea level to reach some of the most rugged and inaccessible peaks these islands have to offer. In fact, some of the summits were so rugged that they were not climbed (at least in modern record) until the ascents of the 1870s when Sheriff Alexander Nicolson, Norman Collie and John MacKenzie climbed them. To commemorate them, three peaks were given commemorative names in Gaelic—Sgurr Thormoid being Norman's Peak, Sgurr MhicChoinnich being MacKenzie's Peak and Sgurr Alasdair being Alexander's peak.

The Black Cuillin form a ring of gabbro and basalt around Loch Coruisk. The traverse is very difficult, and takes most of a summer's daylight hours to complete. Most climbers will make a series of ascents up easier routes in stony corries and gradually tick off the mountains. The southernmost summit is Gars-bheinn, followed by Sgurr a' Choire Bhig and Sgurr nan Eag. Sgurr Dubh Mór is a prominent Munro, with Sgurr Dubh an Da Bheinn being a subsidiary summit of it. Four important Munros follow—with Sgurr Alasdair being the highest point on the ridge, just short of 1,000 metres above sea level. Its first recorded ascent took place in 1873.

Sgurr Dearg has the most inaccessible summit in Britain—An Stac, or the Inaccessible Pinnacle as it is better known. Rock climbing skills are required to reach the summit of this natural monolith which projects 150 feet above the more substantial summit of the main part of the mountain. A very airy ascent, climbers often choose to abseil back down from the summit. The stac was first ascended by Charles and Lawrence Pilkington in 1880, and in 2014 the mountain biker, Danny MacAskill, took his cycle to the top for a popular film.

Sgurr na Banachdaich is another Munro,

dwarfing the smaller Sgurr Thormaid. Sgurr a' Ghreadaidh and Sgurr a' Mhadaidh follow, two more Munros. The ridge drops below 3,000 feet for a stretch, before rising again to Bruach na Frithe and Am Basteir. The former is one of the easiest Cuillin summits to climb, a popular ascent via Fionn Choire. Am Basteir, in contrast, forms one of the more difficult summits to reach, especially if the Bad Step is climbed, graded severe, with vertical rocks below. Sgurr nan Gillean is the last Munro on the ridge.

Most of the mountains in the south of Skye are protected as part of the Cuillin Special Protection area. The main Cuillin summits are the property of the chiefs of the Clan MacLeod, whose seat is Dunvegan Castle, located on the island.

To the east of the Cuillin proper is another range of mountains, referred to as the Red Cuillin. The ridge, though less rugged than the Cuillin, provides a considerable day's climbing. The mountains are separated more, running from Glamaig in the north through Marsco and Garbh-bheinn to Blabheinn, a solitary Munro. This ridge includes Skye's only two Corbetts. Strathaird, Torrin and Sconser estates are protected by the John Muir Trust.

Beinn Dearg Mhór and Beinn na Caillich form a pair of mountains usually ascended together. A second Beinn na Caillich is located at the east end of the island, adjoining the taller Sgurr na Coinneach. Beinn Aslaig to the south is one of those summits which may, or may not, be a mountain, the 2,000 foot contour being of some dubiety. Currently the summit is excluded as a mountain.

The remaining Skye mountains are located to the north in Trotternish. The Storr, with its distinctive rock formations on the east side, including the Old Man of Storr, is the tallest. Stretching north from it are three other mountains and lesser summits, forming a remote ridge walk. Creag a' Lain is another summit that just misses out on inclusion in this list, whereas the farther summit of Beinn Edra just makes it.

The Mountains of Great Britain

Height	Name	NGR	OS L	OS E	Ascent
	Trotternish:				
611	Beinn Edra	NG 455627	23	E408	
639	Baca Ruadh	NG 474575	23	E408/E409	
669	Hartaval	NG 480551	23	E408/E409	
719	The Storr	NG 495541	23	E408/E409	
	Cuillin:				
895	Gars-bheinn	NG 468187	32	E411	
875	Sgurr a' Choire Bhig	NG 466191	32	E411	
924	▲ Sgurr nan Eag	NG 457195	32	E411	
733	Sgurr Dubh Beag	NG 466204	32	E411	
944	▲ Sgurr Dubh Mór	NG 457206	32	E411	
938	Sgurr Dubh an Da Bheinn	NG 455204	32	E411	
947	Sgurr Sgumain	NG 448206	32	E411	
992	▲ Sgurr Alasdair	NG 450208	32	E411	
948	▲ Sgurr Mhic Coinnich	NG 450210	32	E411	
986	▲ An Stac na Sgurr Dearg	NG 445215	32	E411	

The Mountains of Great Britain

Height	Name	NGR	OS L	OS E	Ascent
965	▲ Sgurr na Banachdaich	NG 441225	32	E411	
926	Sgurr Thormaid	NG 441226	32	E411	
973	▲ Sgurr a' Ghreadaidh	NG 445230	32	E411	
918	▲ Sgurr a' Mhadaidh	NG 447235	32	E411	
890 est	Stob Coir' a' Mhadaidh	NG 449236	32	E411	
896	Sgurr Bealach na Glaic Mòire	NG 451237	32	E411	
881	Sgurr Thuilm	NG 439242	32	E411	
688	Sgurr an Fheadain	NG 452245	32	E411	
869	Bidean Druim nan Ramh	NG 456240	32	E411	
860	Sgurr na Bairnich	NG 461244	32	E411	
958	▲ Bruach na Frithe	NG 461252	32	E411	
930	Sgurr a' Fhionn Choire	NG 464252	32	E411	
898	Sgurr a' Bhasteir	NG 464258	32	E411	
934	▲ Am Basteir	NG 466253	32	E411	
964	▲ Sgurr nan Gillean	NG 471253	32	E411	
914	An Stac Mór nan Gillean (Knight's Peak)	NG 472254	32	E411	

736	Sgurr na h-Uamha	NG 476240	32	E411
673	An Coileach	NG 524305	32	E410/E411
775	△Sgurr Mhairi (Glamaig)	NG 513300	32	E410/E411
731	Beinn Dearg Mhór	NG 520285	32	E411
651	Beinn Dearg Mheadhonach	NG 515271	32	E411
736	Marsco	NG 508252	32	E411
702	Belig	NG 544241	32	E411
808	△Garbh-bheinn	NG 531232	32	E411
720	Sgurr nan Each	NG 537227	32	E411
786	Clach Glas	NG 534221	32	E411
929	▲Bla Bheinn (Blabheinn)	NG 530217	32	E411
926	Sgurr Uilleam	NG 529214	32	E411
624	Slat Bheinn	NG 534209	32	E411
709	Beinn Dearg Mór	NG 587229	32	E411/E412
732	Beinn na Caillich	NG 801233	32	E411/E412
732	Beinn na Caillich	NG 771229	33	E412
739	Sgurr na Coinnich	NG 762222	33	E412

22

Rum

For many years referred to as the 'Forbidden Isle' due to its landowners preventing outsiders from landing there, Rum now has a more open access policy. The island is owned by the Nature Conservancy Council, now NatureScot, and the whole island forms a National Nature Reserve, noted for its white-tailed sea eagles, red deer and Manx shearwaters, being one of the largest breeding colonies in the world.

All of the mountains over 2,000 feet on Rum are located on the south-east end of the island, rising steeply from the Sound of Rum. The rock on these summits is mainly gabbro. The northernmost peak is Hallival, a prominent pyramidal mountain that can be reached by following a path from Kinloch into Coire Dubh from where a ridge leads to the summit cairn. Views back over Loch Scresort and north to the Cuillin of Skye are tremendous.

The mountains of Rum are sometimes referred to as the Rum Cuillin, but they do not officially have that name. From Hallival a narrow ridge leads on to Askival, the highest peak on Rum, and one of the island's two Corbetts. It is a prominent summit, dropping steeply on all sides over layered rocks. Views from here are usually made over An Sgúrr of Eigg to the mainland around Moidart and Ardnamurchan.

A steep drop to Bealach an Oir and a sharp climb leads back up to Trollabhal, sometimes referred to as Trallval. Continuing the ridge walk, again the descent and climb back up to Ainshval is very steep, the rocky slopes making rough walking. Ainshval is the second Rum Corbett.

Further south are two more summits, both really part of the same mountain massif, but the northern summit of Leac a' Chaisteil has a re-ascent of over 130 feet, making it eligible for inclusion. The more southerly of the two tops, Sgúrr nan Gillean or Sgór nan Gillean, is taller, the island's southernmost mountain.

Height	Name	NGR	OS L	OS E	Ascent
☐ 722	Hallival (Ailbe-mheall)	NM 395963	39	E397	
☐ 812	△ Askival (Aisge-mheall)	NM 393952	39	E397	
☐ 702	Trollabhal (Trallval)	NM 377952	39	E397	
☐ 781	△ Ainshval (Ais-mheall)	NM 378943	39	E397	
☐ 759	Leac a' Chaisteal (Beinn Mór)	NM 376936	39	E397	
☐ 764	Sgurr nan Gillean	NM 380930	39	E397	

23

Western Isles

Most of the mountains in the western Isles are located on Harris, more accurately on North Harris, where a high stretch of mountains extends north-westwards from Tarbert. The tallest of these is An Cliseam, often Anglicised to Clisham, the Western Isles' only Corbett. Various routes can be made to the top from the road linking Tarbert with Stornoway. The summit rises north of the A859, from which it can be climbed alongside the Allt Tomnabhal. The mountain is covered with wild thyme.

West of An Cliseam are three mountains arranged on a north-south axis, Mullach an Langa at the northern end, a rounded summit that just qualifies as a mountain. The other two summits are Mulla bho Thuath and Mull bho Dheas, the southernmost being the taller of the two. An ascent of the four summits is sometimes referred to as the Clisham Horseshoe.

The deep glen of Bealach a' Sgáil separates the three mountains with another two to the west. Of these, Uisgneabhal Mór is the tallest, a prominent peak when viewed from the south. Behind it and similarly steep is Téileasbhal.

Gleann Mhiabhaig separates a further two mountains to the west, Oireabhal and Ulabhal. Although it doesn't affect the ascent of Ulabhal, the great rocks of Srón Uladail to the north are worth seeing, reckoned to form the tallest overhang in Britain. Gleann Chliostair with its two lochs separate Oireabhal from Tiorga Mór to the west. A prominent peak, this mountain is often climbed from Loch Chliostair dam.

On South Uist is another lonely mountain, Beinn Mhór, located on the east side of the island in a hilly stretch of countryside. Beinn Mhór is a long ridge of rock running north-south, the highest point a steep ascent from Loch Aineort to the south. Hecla, or Thacla, to the north unfortunately misses being a mountain by twelve feet.

Height	Name	NGR	OS L	OS E	Ascent
679	Tiorga Mór (Tirga Mór)	NB 056116	13/14	E456	
659	Ulabhal (Ullaval)	NB 086115	13/14	E456	
662	Oireabhal (Oreval)	NB 084100	13/14	E456	
697	Teileasbhal (Teilesval)	NB 126091	13/14	E456	
729	Uisgneabhal Mór (Uisgnaval Mór)	NB 121086	13/14	E456	
614	Mullach na Langa	NB 143095	13/14	E456	
720	Mulla bho Thuath	NB 140084	13/14	E456	
743	Mulla bho Dheas	NB 133066	13/14	E456	
799	△An Cliseam (Clisham)	NB 1405073	13/14	E456	
620	Beinn Mhór (Buail' a' Ghoill)	NF 809311	22	E453	

24

Mull

Most of the central and southern part of Mull is covered with substantial hills, one of which reaches Munro height, the only island Munro outwith Skye. Beinn Mhór, or Ben More as it has been Anglified, is the furthest west of Mull's mountains, reaching 3,169 feet. The most popular route to climb the mountain is from Dhiseig, on Loch na Keal, from where a footpath ascends from sea level to the summit. An attractive but tricky rocky ridge leads east to A' Chioch, and keen walkers will probably wish to make a return to Dhiseig over the summit of Beinn Fhada and its long north-western ridge.

On the other side of the bealach in which the Cárn Cul Righ Albainn is located, rise a pair of attractive summits—Cruachan Dearg and Corra-bheinn, both a similar height. They are usually climbed together. Beyond Beinn a' Mheadhain, which just fails to reach mountain status, is a fourth summit, Cruach Choireadail, which rises over the 2,000 foot contour by a few feet.

Beinn Talaidh rises above its near neighbours between Glen More and Loch Bá. It is most easily climbed from near Torness in Glen More.

Between Craignure and Strathcoil rise a group of hills, of which Dún da Ghaoithe is the highest, counting as a Corbett, the only one on Mull. Flanking it is Mainnir nam Fiadh, Beinn Thunacaraidh and Beann Mheadhan. Beinn Thunacaraidh is just over the cusp of qualifying as a separate mountain. Sgurr Dearg lies south of this range.

The final two mountains on Mull are located above Lochbuie, standing to either side of Gleann a' Chaiginn Mhóir. Beinn Buidhe (or Ben Buie) is the taller of the two, rising steeply above the loch, the highest of the triple summits being that to the south. Creach-Beinn, like Beinn Buidhe, is usually climbed from the south, but both are just as easily scaled from Glen More to the north.

The Mountains of Great Britain

Height	Name	NGR	OS L	OS E	Ascent
☐ 702	Beinn Fhada	NM 540349	47/48	E375	
☐ 867	A' Chioch	NM 534334	47/48	E375	
☐ 966	▲Beinn Mhòr (Ben More)	NM 526331	47/48	E375	
☐ 704	Cruachan Dearg	NM 568332	47/48	E375	
☐ 704	Corra-bheinn	NM 573322	48	E375	
☐ 618	Cruach Choireadail	NM 595305	48	E375	
☐ 763	Beinn Talaidh	NM 625347	49	E375	
☐ 637	Beinn Mheadhan	NM 654379	49	E375	
☐ 648	Beinn Thunacaraidh	NM 661369	49	E375	
☐ 766	△Dun da Ghaoithe	NM 672362	49	E375	
☐ 757	Mainnir nam Fiadh	NM 676353	49	E375	
☐ 741	Sgurr Dearg	NM 665340	49	E375	
☐ 698	Creach-Bheinn	NM 643276	49	E375	
☐ 717	Beinn Buidhe (Ben Buie)	NM 604271	49	E375	

25

Jura

The island of Jura only has three summits that reach 2,000 feet, but their shape and name has given them greater fame than many other hills of a similar size. The hills are known as the Paps of Jura, 'pap' being an old Norse word for breast, but strangely there are three of them. Older accounts of the area state that only Beinn a' Chaolais and Beinn an Oir are regarded as the two paps. Scots Gaelic name them as Sgurr na Ciche. Their distinctive shape makes them visible from many miles away.

Most hillwalkers will make an ascent of the three summits in a single expedition. Access to the foot of the hills can be made more readily by following tracks from two possible starts. From Feolin a track northwards to Inver Cottage and Cnocbreac is followed up the western slopes below Beinn a' Chaolais, the westmost pap. Like the other summits, the hill comprises quartzite rock.

To the north-east of the first pap rises Beinn an Oir, which merits inclusion in Corbett's list. Its name translates as mountain of gold. One of the earliest recorded ascents took place around 1772, when Thomas Pennant climbed it. The rock is layered, but is angled, and the slopes are often covered with scree, especially on the steep western flanks, overlooking Na Garbh-lochanan. In the late eighteenth century experiments were undertaken near the summit as to how altitude affected the boiling temperature of water. A few ruinous huts connected with this survive.

The third pap is located east of Beinn an Oir, Beinn Shiantaidh. This summit is just short of 2,500 feet, otherwise it would have been a Corbett too. To the south of this mountain is Loch an t-Síob', a track to the boathouse of which makes an easier approach to the paps from the east side of Jura.

The Jura Fell Race takes in the three paps in a long race which includes four other summits in the sixteen-mile route.

Height		Name	NGR	OS L	OS E	Ascent		
733	☐	Beinn a' Chaolais	NR 489735	60/61	E355			
785	☐ △	Beinn an Oir	NR 498750	60/61	E355			
757	☐	Beinn Shiantaidh	NR 513749	61	E355			

26

Glen Coe & Glen Creran

Glen Coe is one of Scotland's most iconic glens, the steep and shapely peaks on either side of the narrow valley being readily identified by many climbers, and the glen's bloody history adds to its character. It was here, in 1692, that many of the MacDonald residents were massacred by their Campbell guests, acting on behalf of the Scottish government.

On the north side of the glen is a ridge of substantial peaks, with the Aonach Eagach ridge forming one of the most challenging scrambles for the average hill-walker. On this ridge are two Munros, Sgórr nam Fiannaidh and Meall Dearg, plus two Munro tops, Stob Coire Leith and Am Bodach. Most climbers just make an ascent of these four tops, whereas the longer ridge extends from Sgórr na Ciche, or the Pap of Glencoe to the west (a distinctive cone in many views of Loch Leven) to Beinn Bheag to the east. The latter mountain is rather low and indistinct when compared with the greater summits in the district. A few other summits rise to the east of the head of Glen Coe, including Beinn a' Chrulaiste, a Corbett, which has a spectacular view west down through the Pass of Glencoe. North of the range is a second Corbett, Garbh Bheinn, which rises above Kinlochleven, and where in fiction David Balfor hid in Stevenson's *Kidnapped*.

A fair amount of the northern side of Glen Coe is owned by the National Trust for Scotland, whereas most of the southern side is Trust property, including the great sentinels of Buachaille Etive Mór and Buachaille Etive Beag. The lands were acquired in 1937 after an appeal by Percy Unna, who intended that the 'land should be maintained in its primitive condition for all time'. Each of the Buachaille ridges contains two Munros, having names of their own, Stob Dearg in the case of Etive Mór and Stob Dubh in Beag's case. On Buachaille Etive Mór, Stob na Bróige was promoted to Munro status in 1997, as was Stob Coire Raineach on

Etive Beag. The front of Stob Dearg is one of Scotland's greatest iconic views, seen across the flat extent of Rannoch Moor.

Also Trust property are the northern slopes of Bidean nam Bian, a substantial Munro of 3,766 feet which has five other tops listed by Munro on its flanks, including the attractive Stob Coire nan Lochan and the Three Sisters, which abut onto Glen Coe. This mountain is the highest in Argyll. On the south side is Beinn Maol Chaluim, a Corbett.

Beyond the Fionn Ghleann are a couple of mountains over 3,000 feet, Sgúrr na h-Ulaidh being a Munro. It, Beinn Fhionnlaidh and Beinn Sgulaird are three quieter and less rugged Munros. East of Beinn Sgulaird rises Beinn Trilleachan, a rocky mountain rising on the west side of the head of Loch Etive, which qualifies as a Corbett. It is thought few mountains outside Skye have so much exposed rock than this summit.

In the land of Appin the lower slopes are generally afforested, and rising above the woods is Beinn a' Bheithir. The highest point is Sgórr Dhearg, at 3,361 feet, but the slightly lower Sgórr Dhonuill to the west qualifies as a Munro also. Also in the vicinity and surrounded by forestry is Fraochaidh, a Corbett, and to its east is a second, Meall Lighiche.

Creach Bheinn is another Corbett that rises to the east of Barcaldine, usually climbed from Druimavuic at the head of Loch Creran. Lower mountains continue to the south, with Beinn Bhreac and its large summit cairn above Barcaldine and Beinn Mheadhonach near to Bonawe.

In addition to the extensive Glen Coe and Dalness estate of the National Trust for Scotland, the west side of Glen Etive is part of the private Blackmount Estate. Other estates include Altnafeadh, Glenure and Ardchattan. Much of the remaining countryside is part of Forestry and Land Scotland. Ben Nevis and Glencoe National Scenic Area includes much of Glen Coe and surrounding areas, and the glen itself is a national nature reserve, established in 2017.

27

Height	Name	NGR	OS L	OS E	Ascent
739	Stob na Cruaiche (A' Chruach)	NN 363571	41	E385	
647	Meall nan Ruadhag	NN 298577	41	E385	
705	Meall Bhalach Ear	NN 269572	41	E384	
708	Meall Bhalach Iar	NN 260577	41	E384	
857	△ Beinn a' Chrulaiste	NN 246567	41	E384	
616	Beinn Bheag	NN 221579	41	E384	
707	Stob Mhic Mhartuin	NN 208576	41	E384	
	Aonach Eagach:				
903	A' Chailleach	NN 189579	41	E384	
873	Srón Gharbh	NN 178585	41	E384	
943	Am Bodach	NN 168580	41	E384	
953	▲ Meall Dearg	NN 161583	41	E384	
940	Stob Coire Leith	NN 149585	41	E384	
967	▲ Sgórr nam Fiannaidh	NN 125594	41	E384	
742	Sgórr na Ciche (Pap of Glencoe)	NN 125594	41	E384	
867	△ Garbh Bheinn	NN 169601	41	E384	

27

Royal Forest:

1022	☐	▲ Stob Dearg - Buachaille Etive Mór	NN 222542	41	E384
902	☐	Stob Coire na Tulaich	NN 214342	41	E384
1011	☐	Stob na Doire	NN 208532	41	E384
941	☐	Stob Coire Altruim	NN 198531	41	E384
956	☐	▲ Stob na Broige	NN 191525	41	E384
925	☐	▲ Stob Coire Raineach - Buachaille Etive Beag	NN 191548	41	E384
958	☐	▲ Stob Dubh	NN 179535	41	E384
778	☐	Cárn Coire nan Easan	NN 166528	41	E384
1072	☐	▲ Stob Coire Sgreamhach	NN 154536	41	E384
952	☐	Beinn Fhada	NN 159540	41	E384
931	☐	Beinn Fhada - Centre Top	NN 164543	41	E384
823	☐	Beinn Fhada - N.E. Top	NN 170550	41	E384
892	☐	Aonach Dubh	NN 150559	41	E384
1115	☐	Stob Coire nan Lochan	NN 149549	41	E384
1150	☐	▲ Bidean nam Bian	NN 141542	41	E384
1107	☐	Stob Coire nam Beith	NN 139546	41	E384

27

Height	Name	NGR	OS L	OS E	Ascent
907	△ Beinn Maol Chaluim	NN 135526	41	E384	
748	Meall a' Bhuiridh	NN 126507	41	E384	
994	▲ Sgòr na h-Ulaidh	NN 111519	41	E384	
968	Stob an Fhuarain	NN 118523	41	E384	
845	Aonach Dubh a' Ghlinne	NN 120533	41	E384	
772	△ Meall Lighiche	NN 094529	41	E384	
678	Meall an Aodainn	NN 080525	41	E384	
663	Sgòrr a' Choise	NN 084551	41	E384	
676	Meall Mòr	NN 106559	41	E384	
	Appin:				
947	Sgòrr Bhan	NN 062561	41	E384	
1024	▲ Sgòrr Dhearg - Beinn a' Bheithir	NN 057558	41	E384	
1001	▲ Sgòrr Dhonuill	NN 040555	41	E384	
824	Mullach Choire Dheirg	NN 027558	41	E384	
758	Creag Ghorm	NN 033578	41	E384	
626	Càrn Beag	NN 054534	41	E384	

27

718	Fraochaidh Ear	NN 049522	41	E384
879	△Fraochaidh	NN 029517	41	E384
627	Beinn Mhic na Ceisich	NN 016493	50	E384
654	Meall Ban	NM 997499	49	E376/E384
959	▲Beinn Fhionnlaidh	NN 095498	50	E384
841	Stob Coire na Tullaich	NN 106498	50	E384
690	Stob Gaibhre	NN 063467	50	E384
937	▲Beinn Sgulaird	NN 053461	50	E384
909	Stob Coire nan Tulach	NN 055463	50	E384
848	Meall Garbh	NN 047452	50	E384
863	Beinn Sgulaird Deas	NN 043447	50	E377
662	Creag na Cathaig	NN 037434	50	E377
804	Meall Garbh	NN 029429	50	E377
810	△Creach Bheinn	NN 024422	50	E377
726	Beinn Bhreac Mór	NN 008409	50	E376/E377
708	Beinn Bhreac	NM 993400	49	E376/E377
690	Beinn Molurgainn	NN 019400	50	E377

27

27

Height	Name	NGR	OS L	OS E	Ascent
☐ 715	Beinn Mheadhonach	NN 020369	50	E377	
☐ 767	Meall nan Gobhar	NN 096447	50	E384	
☐ 840	△ Beinn Trilleachan	NN 087439	50	E377	

Monadh Dubh & Lorn

The Black Mount, or Am Monadh Dubh as it is called in Gaelic, is a well-known set of mountains when viewed across Rannoch Moor and from the east end of Glen Coe. The Glencoe Mountain Resort is located off the A82, on the northern slopes of Coire Pollach, with the famous Blackrock Cottage nearby, so often seen on calendars. The ski centre was Scotland's first, being established in 1960, and has ski tows and bike trails on the slopes of Meall a' Bhuiridh, the northernmost Munro in this section.

Immediately to the west are the mountains of Creise and Clach Leathad, the former the highest by just one metre, though Clach Leathad (sometimes spelt Clachlet) is perhaps the better known. West of this pair of mountains is Beinn Mhic Chasgaig, a Corbett summit.

To the south rises a group of mountains and ridges with Stob Ghabhar on the boundary line, one of the most dominant mountains in the area. It has various ridges off the top, encircling a fine selection of corries. On one of these is Stob a' Bhruaich Leith. Stob a' Choire Odhair rises to the east, a separate Munro. To the west are a series of mountains as far as Glen Etive. Stob Dubh rises steeply from almost sea-level, the Corbett summit a steady pull over granite rocks.

The next block of mountains (south of the Gleann Ceitlin—Loch Dochart gap), is dominated by Stob Coir' an Albannaich. On the top is a large cairn, originally used by the Ordnance Survey when making early maps, the view from here being considerable on clear days. Meall nan Eun to the east qualifies as a Munro, but its lower height means that it is dwarfed by both Albannaich and Ghabhar.

The ridge over Glas Bheinn Mhór and Meall nan Tri Tighearnan leads to Beinn Starabh (or Beinn Starbhanach), another of the mountains in this section that rises over the 1,000 metre line. Ascents are usually

28

made from the head of Loch Etive, the two previously listed hills often included in a circle of Coire Dearg and Coire Dubh.

Beinn nan Aighenan is a remote Munro, requiring a long trek across country to reach. A common means of ascent is to include the summit in a long outing from Glen Etive that includes Beinn Starabh. Ascents from Forest Lodge at Loch Tulla involve lengthy walks in.

The next range of summits to the south extends from Beinn a' Chochuill in the south-west to Beinn Suidhe in the north-east. Beinn a Chochuill and its neighbour Beinn Eunaich are Munros, usually ascended from Castles to the south, the hydro road assisting access up towards Lairig Noe. The mountains stretching north-east from here do not reach Corbett status, meaning that they are much quieter and rarely climbed, and yet Meall Garbh above Glenkinglass Lodge is an attractive peak.

The Cruachan Beann range rises steeply above the Pass of Brander, where the A85 makes its way westwards between the mountains and the north end of Loch Awe. There are a number of peaks in the Cruachan range that merit inclusion as separate mountains, the highest being Ben Cruachan itself, which is over 1,100 metres in height. Stob Daimh counts as a Munro in itself, and the south-eastern summit, Beinn a' Bhúiridh, is a Corbett.

Ascents of the Cruachan range often include a walk to the Cruachan Reservoir, created as part of the hydro-electric scheme which pumps water from Loch Awe up to the reservoir and then releases it through the subterranean power station, carved into the mountain. The power station has a capacity of 440 MW and was open by Queen Elizabeth in 1965.

The hills of Glen Orchy are blanketed with forestry plantations on their lower levels in many places, making access more difficult, although some forest roads help. Beinn Mhic Mhonaidh and Beinn Udlaidh are Corbetts, resulting in them being more popular climbs than the lower heights hereabouts.

28

The Mountains of Great Britain

Ascent

Height	Name	NGR	OS L	OS E	Ascent
	Monadh Dubh:				
1108	▲Meall a' Bhuiridh	NN 251503	41	E384	
1100	▲Creise	NN 239507	41	E384	
1099	Clach Leathad	NN 240493	50	E384	
864	△Beinn Mhic Chasgaig	NN 221502	41	E384	
834	Beinn Toaig	NN 262455	50	E377/E384	
945	▲Stob a' Choire Odhair	NN 257460	50	E377/E384	
867	Aonach Mór	NN 219479	50	E377/E384	
774	Creag a' Bhealaich	NN 228439	50	E377/E384	
1090	▲Stob Ghabhar	NN 230455	50	E377/E384	
697	Meall an Araich	NN 218436	50	E377	
941	Stob a' Bhruaich Leith	NN 209460	50	E377/E384	
890	Meall Odhar	NN 195464	50	E377/E384	
703	Meall Garbh	NN 197481	50	E384	
845	Beinn Ceitlin	NN 174488	50	E384	
883	△Stob Dubh	NN 167488	50	E384	

The Mountains of Great Britain

Height	Name	NGR	OS L	OS E	Ascent										
928	▲ Meall nan Eun	NN 192449	50	E377											
877	Meall Tarsuinn	NN 180449	50	E377											
1044	▲ Stob Coir' an Albannaich	NN 169443	50	E377											
997	▲ Glas Bheinn Mhór	NN 153429	50	E377											
892	Meall nan Tri Tighearnan	NN 145425	50	E377											
1078	▲ Beinn Starabh	NN 126427	50	E377											
930	Meall Cruidh	NN 129416	50	E377											
918	Stob an Duine Ruaidh	NN 124410	50	E377											
709	Beinn nan Lus	NN 131375	50	E377											
957	▲ Beinn nan Aighenan	NN 149404	50	E377											
744	Am Binnein	NN 165407	50	E377											
639	Beinn Inbhirbheigh	NN 272382	50	E377											
665	Meall Tairbh	NN 252377	50	E377											
676	Beinn Suidhe	NN 212400	50	E377											
624	Meall Buidhe	NN 183376	50	E377											
701	Meall Garbh	NN 167367	50	E377											

The Mountains of Great Britain

		Height	Name	Grid Ref		
☐		693	Meall Beithe	NN 169351	50	E377
☐		719	Beinn Lurachan	NN 167339	50	E377
☐		810	Meall Copagach	NN 153341	50	E377
☐		880	Stob Coire Easan	NN 144334	50	E377
☐		988	▲ Beinn Eunaich	NN 136328	50	E377
☐		980	▲ Beinn a' Chochuill	NN 118327	50	E377
			Cruachan Beann:			
☐		897	△ Beinn a' Bhuiridh	NN 094283	50	E360/E377
☐		980	Stob Garbh	NN 094303	50	E377
☐		966	Srón an Isean	NN 099311	50	E377
☐		998	▲ Stob Daimh	NN 094308	50	E377
☐		1024	Drochaid Ghlas	NN 084305	50	E377
☐		916	Meall Cuanail	NN 069295	50	E377
☐		1126	▲ Ben Cruachan (Cruachan Beann)	NN 069305	50	E37
☐		1101	Stob Dearg	NN 063307	50	E377
☐		906	Meall nan Each	NN 055316	50	E377

28

Height	Name	NGR	OS L	OS E	Ascent
	Glen Orchy:				
644	Na Cruachan	NN 190312	50	E39/E377	
650	Beinn Donachain	NN 195314	50	E39/E377	
796	△Beinn Mhic-Mhonaidh	NN 208350	50	E377	
679	Meall Aluinn	NN 221355	50	E377	
636	Beinn na Sroine	NN 233289	50	E39/E377	
840	△Beinn Udlaidh	NN 280332	50	E377	
802	Beinn Bhreac-liath	NN 303340	50	E377	
653	Beinn Bheag	NN 316326	50	E377	

Glen Lyon & Mamlorn

Stretching from the Loch Rannoch-Loch Tummel valley south to Loch Tay and Glen Dochart, this section contains a great range of mountains forming the backbone of Breadalbane. The eastmost hills lie north of Aberfeldy, only the two Corbetts being of any significance—Meall Tairneachan and Farragon Hill.

West of Glengoulandie is the great peak of Schiehallion, the 'fairy hill of the Caledonians' which appears conical when viewed end on. This natural feature made it ideal for early experiments in calculating the mass of the Earth, carried out in 1772 by Nevil Makeleyne, the Astronomer Royal.

On the north side of eastern Glen Lyon is a range of mountains containing four Munros. Meall nan Aighean lies furthest east, and Cárn Mairg is between it and Schiehallion. West of here is Meall Garbh and Cárn Gorm. The next group of hills are not so tall, but among them are Beinn Dearg and Cam Chreag (both Corbetts). North of Loch an Daimh rises Meall Buidhe, a lonely Munro, its neighbour, Garbh Mheall, probably a more interesting summit, but which fails to reach 3,000 feet by a few feet.

On the southern side of Loch an Daimh rises Stúc an Lochain, an attractive Munro with a deep corrie down into Lochan nan Cat. A few outlying mountains form the shoulders of this summit, including An Grianan, the rocky slopes of which rise above Glen Lyon. At the head of Loch an Daimh is Srón a' Choire Chnapanaich, a Corbett, and west of it is the lonely Meall Buidhe, a Corbett that almost reaches Munro status.

At the western end of this section is a range of considerable Munros that are fairly well known. Beinn a' Chreachain is one of the highest, its lofty peak rising above steep Coire an Lochain. A number of Munro tops surround it, followed by Beinn Achaladair, the summit of which affords extensive panoramas over Rannoch Moor to the north.

29

Beinn an Dothaidh and Beinn Dorain are two great bulks of mountains rising high above the busy road heading through Bridge of Orchy. They are often climbed as a pair, a steep path from the station heading to the bealach from where two separate ascents are often made. Beinn Dórain is famed in Gaelic song. To the east, above the headwaters of Loch Lyon, is Beinn a' Chuirn and Beinn Mhanach, the latter a lonely Munro.

Between glens Dochart and Lochay is a range of mountains containing a couple of Munros—Sgiath Chuil and Meall Glas. A few other summits rise over 3,000 feet, and to the west of the range is Beinn nan Imearean, a Corbett. These hills form the northern boundary of the Loch Lomond and Trossachs National Park. So, too, does Beinn Challum, the Stob Glas summit of the twin peaked mountain being the highest.

To the west of Beinn Challum, and rising above Tyndrum to Loch Lyon, are five great mountains that count as Corbetts, indicating their individual prominence. These are Beinn Chaorach, Beinn Odhar, Beinn a' Chaisteal, Beinn nam Fuaran and Cam Chreag.

On the north side of Glen Lochay can be found Creag Mhór and Beinn Sheasgarnaich, two Munros in excess of 1,000 metres. Further east rises Meall Ghaordaidh.

To the north of Killin is a group of mountains often referred to as the Tarmachans, after the tallest—Meall nan Tarmachan, which rises to 3,421 feet above sea level. Meall Garbh, the two summits of Beinn nan Eachan and Creag na Caillich are Munro tops in the group. To the north of here, on opposite sides of the minor road that crosses from Loch Tay to Glen Lyon, are two Corbetts, Beinn nan Oighreag and Meall nam Maigheach.

Eastwards is the Ben Lawers group, this mountain being almost 4,000 feet tall—indeed there was at one time a plan to build a huge cairn to bring it over the magic contour.

Height	Name	NGR	OS L	OS E
1083	▲ Schiehallion (Sidh Chaillean)	NN 713548	42/51/52	E49
822	Creag Mhór	NN 712490	51/52	E48/E49
904	Cárn Mhic Ghriogair	NN 706495	51/52	E48/E49
981	▲ Meall nan Aighean	NN 694497	51	E48/E49
874	Meall nan Eun	NN 708509	42/51/52	E48/E49
1012	Meall Liath	NN 692512	42/51	E48/E49
1041	▲ Cárn Mairg	NN 684513	42/51	E48/E49
792	Geal Charn	NN 681545	42/51	E49
1004	Meall a' Bharr	NN 671516	42/51	E48/E49
802	Meall Breac	NN 638542	42/51	E49
968	▲ Meall Garbh	NN 649515	42/51	E48/E49
924	An Sgórr	NN 641510	42/51	E48/E49
1029	▲ Cárn Gorm	NN 634501	42/51	E48/E49
830	△ Beinn Dearg	NN 609497	51	E48/E49
741	Creag Ard	NN 601488	51	E48/E49
745	Meall a' Bhuic	NN 580508	42/51	E48/E385

Ascent

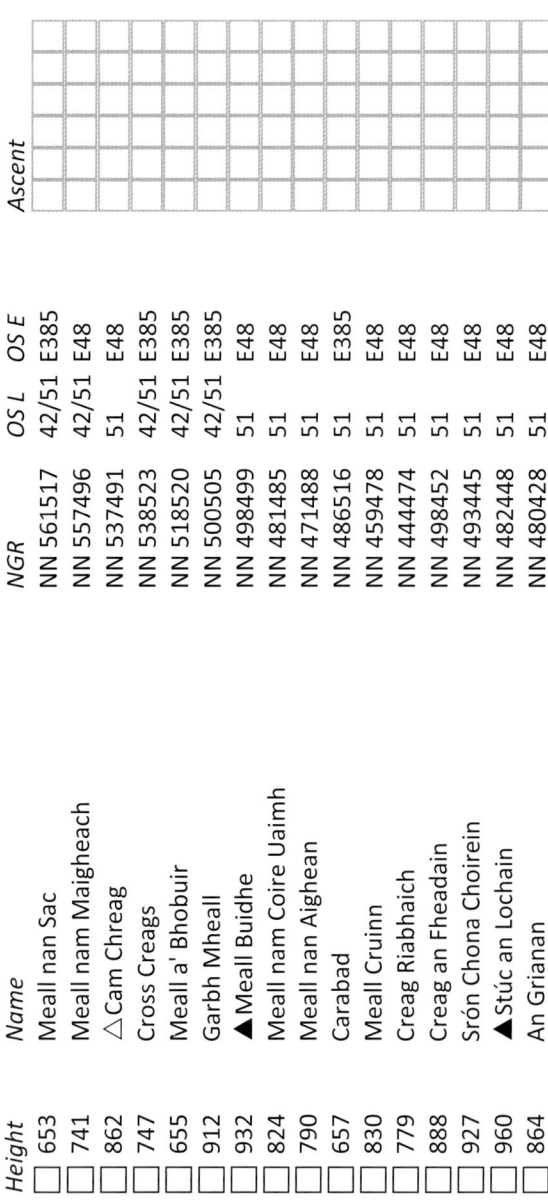

Height	Name	NGR	OS L	OS E	Ascent
653	Meall nan Sac	NN 561517	42/51	E385	
741	Meall nam Maigheach	NN 557496	42/51	E48	
862	△Cam Chreag	NN 537491	51	E48	
747	Cross Creags	NN 538523	42/51	E385	
655	Meall a' Bhobuir	NN 518520	42/51	E385	
912	Garbh Mheall	NN 500505	42/51	E385	
932	▲Meall Buidhe	NN 498499	51	E48	
824	Meall nam Coire Uaimh	NN 481485	51	E48	
790	Meall nan Aighean	NN 471488	51	E48	
657	Carabad	NN 486516	51	E385	
830	Meall Cruinn	NN 459478	51	E48	
779	Creag Riabhaich	NN 444474	51	E48	
888	Creag an Fheadain	NN 498452	51	E48	
927	Srón Chona Choirein	NN 493445	51	E48	
960	▲Stúc an Lochain	NN 482448	51	E48	
864	An Grianan	NN 480428	51	E48	

The Mountains of Great Britain

815	Meallan Odhar	NN 471448	51	E48
837	△ Srón a' Choire Chnapanaich	NN 456452	51	E48
798	Creag a' Chaorainn	NN 442438	51	E48
910	△ Meall Buidhe	NN 427450	51	E48
874	Meall Dail	NN 411435	51	E48
831 est	Meall na Feithe Faide	NN 412452	51	E48
800	Guala Mhór	NN 403458	51	E48
1081	▲ Beinn a' Chreachain	NN 373441	50	E48/E377
978	Meall Buidhe	NN 365443	50	E377
1036	▲ Beinn Achaladair	NN 343432	50	E377
1002	Beinn Achaladair Deas	NN 342421	50	E377
1004	▲ Beinn an Dothaidh	NN 328409	50	E377
1076	▲ Beinn Dorain	NN 326379	50	E377
923	Beinn a' Chuirn	NN 360411	50	E377
953	▲ Beinn Mhanach	NN 373412	50	E48/E377
719	Creag Mhór	NN 513340	51	E48
796	Beinn Bhreac	NN 501329	51	E48

189

Height	Name	NGR	OS L	OS E	Ascent
866	Meall na Samhna	NN 489333	51	E48	
883	Cárn Lobhaidh	NN 475331	51	E48	
921	▲ Sgiath Chuil	NN 463318	51	E48	
937	Beinn Cheathaich	NN 444326	51	E48	
959	▲ Meall Glas	NN 432322	51	E48	
849	△ Beinn nan Imirean	NN 419308	51	E48	
998	Beinn Challum Deas	NN 386314	50	E39/E48/E377	
1022	▲ Beinn Challum - Stob Glas	NN 387323	50	E48/E377	
818	△ Beinn Chaorach	NN 359328	50	E377	
653	Meall Buidhe	NN 342319	50	E39/E377	
901	△ Beinn Odhar	NN 337338	50	E377	
883	△ Beinn a' Chaisteil	NN 347364	50	E377	
806	△ Beinn nam Fuaran	NN 361382	50	E377	
884	△ Cam Chreag	NN 375347	50	E48/E377	
956	Stob nan Clach	NN 388351	50	E48/E377	
1047	▲ Creag Mhór	NN 390361	50	E48	

895	Meall Tionail	NN 389377	50	E48
1078	▲ Beinn Sheasgarniach	NN 413383	51	E48
759	Meall a' Chall	NN 436403	51	E48
779	Creag nam Bodach	NN 445377	51	E48
806	△ Meall nan Subh	NN 461397	51	E48
712 est	Srón Eanchainne	NN 481396	51	E48
792	Meall Taurnaigh	NN 488387	51	E48
846	Meall na Cnap Laraich	NN 499398	51	E48
1039	▲ Meall Ghaordaidh	NN 514398	51	E48
809	Creag an Tulabhain	NN 525416	51	E48
909	△ Beinn nan Oighreag	NN 543414	51	E48

Na Tarmachain:

749	Meall Dhuin Croisg	NN 547371	51	E48
815	Meall Ton Eich	NN 554389	51	E48
914	Creag na Caillich	NN 562376	51	E48
1000	Beinn nan Eachan	NN 570384	51	E48
1026	Meall Garbh	NN 579383	51	E48

Height	Name	NGR	OS L	OS E	Ascent
1044	▲Meall nan Tarmachan	NN 585390	51	E48	
923	Càrn a' Mhoirneas	NN 589385	51	E48	
842	Creag an Lochain	NN 591403	51	E48	
642	Creag nan Eildeag	NN 597461	51	E48	
779	△Meall nam Maigheach	NN 585435	51	E48	
926	▲Meall a' Choire Leith	NN 612439	51	E48	
1069	▲Meall Corranaich	NN 616410	51	E48	
1103	▲Beinn Ghlas	NN 626404	51	E48	
1214	▲Beinn Lawers (Beinn Labhar)	NN 634414	51	E48	
1117	▲An Stúc	NN 639431	51	E48	
1123	▲Meall Garbh	NN 644437	51	E48	
1001	▲Meall Greigh	NN 674437	51	E48	
	North Dull:				
654	Creag Chean	NN 795532	42/51/52	E49	
787	△Meall Tairneachan	NN 807544	52	E49	
736	Creag an Loch	NN 822542	52	E49	

Creag a' Mhadaidh	NN 832542	52	E49
△ Farragon Hill	NN 840553	52	E49
Creag an Lochain	NN 843562	52	E49
Cárn na Lice	NN 839567	52	E49
Beinn Eagagach	NN 856566	52	E49
Tom an Fhuarain	NN 863562	52	E49
Meall a' Choire	NN 881568	52	E49
Meall a' Charra	NN 891576	52	E49

679	783	651	656	691	630	620	617
☐	☐	☐	☐	☐	☐	☐	☐

Glens Almond, Cuaich & Lednock

This section of the southern highlands is bounded by the A85 to the south, the A84 to the west, the A827 and Loch Tay to the north and the A9 and River Tay to the east. Much of the hill country is occupied by sporting estates, and the eastern flanks of Glen Ogle are included within the Loch Lomond and Trossachs National Park.

There is only one Munro within the bounds of the area—Beinn Chonzie, which reaches 931m in height. Being over the 3,000 foot line, this summit is a popular climb from either the dam at Loch Turret, or else from Coishavachan at Invergeldie in Glen Lednock. The former route takes one up the most attractive side of the mountain, whereas the latter route is shorter. Although written as Beinn, or Ben Chonzie, the name is pronounced 'Ben-y-Hone', which means either 'mountain of the weeping', or 'mossy hill'. The summit has a shelter, located at the junction of three march fences. It affords extensive panoramas north to Ben Lawers and surrounding hills.

Loch Turret was a natural loch which had a poem written about it by Robert Burns, 'On scaring some waterfowl in Loch Turit, a wild scene among the hills of Oughtertyre'. In it Burns wrote that 'Nature's gifts to all are free'. On the east side of Loch Turret rises Auchnafree Hill, a Corbett, but one which is rather undistinguished. To its south is Cnoc Choinneachain, on the upper slopes of which is Cárn Chainichin (obviously a variation of the spelling) which is reputed to be King Kenneth's Cairn, of date AD 1003.

Rob Roy's Way passes through the hills from Loch Freuchie to upper Glen Almond to Loch Tay. On either side the hills rise steeply from the glens, Creagan na Beinne being the highest on the north side of the valley. This is a remote Corbett, rather bland on the summit, but a steep climb from any side. East of the summit, a ridge extends towards Glen Lochan on which are a series of high summits—Srón a' Chaoineidh, Meall nan Eanchainn and Meall nam Fuaran. Each of these mountains have tracks to their summits, created for shooting parties, but which allow quicker walking to the tops.

South-west of Rob Roy's Way is Creag Uchdag, which rises steeply above Loch Lednock. This too is a Corbett, often climbed from Ardeonaig on Loch Tayside, or from Glen Lednock to the south. On some older maps it is named Creag Uigeach. Loch Lednock was created by the construction of a dam across the narrows above the waterfall known as Sput Rolla, the cascades of which have been reduced due to the diversion of some of the water. These waters are used for generating electricity. The dam is a specially-designed type, extra strong to withstand the small earthquakes that often occur in this area due to the existence of the Highland Boundary Fault.

West of Loch Lednock the hills are lower, but in various places the summits are rugged, such as the rocks of Creag Ruadh, Creag Gharbh and Meal Odhar. Below Creag Gharbh is Lochan Breaclaich, a natural sheet of water extended by the construction of a dam, part of the Lednock power generating scheme, the water passing through a tunnel to Loch Lednock, from where the water drives St Fillans Power Station.

Two summits rise to the east of the Sma' Glen to Amulree road—Meall nan Caorach and Meall Reamhar. The track from Girron, near Amulree, allows both summits to be reached fairly quickly.

North-east of Gleann Cuaich (or Glen Quaich) rise another group of low summits, the highest of which is Meall Dearg, positioned above Glen Cochill. Tracks for grouse shooting make their way near to the summit. To the west is Meall Odhar, much of its northern and western slopes disfigured by roads for the Calliacher Wind Farm. On the western slopes of Meall a' Choire Chreagaich are a couple more turbines.

30

Height	Name	NGR	OS L	OS E	Ascent
690	Meall Dearg	NN 887414	52	E379	
663	Creag an Loch	NN 879408	52	E379	
623	Meall Odhar	NN 856425	52	E379	
619	Creag Choille	NN 856409	52	E379	
631	Tir Eilde	NN 832420	52	E379	
667	Meall Reamhar	NN 876326	52	E379	
704	Meall Mór	NN 849349	52	E379	
685	Creag Grianain	NN 842351	52	E379	
730	Beinn na Gainimh	NN 836344	52	E379	
805	Meall nam Fuaran	NN 826361	52	E379	
797	Meall Tuath nam Fuaran	NN 820370	52	E379	
737	Garrow Hill	NN 812378	52	E379	
858	Meall nan Eanchainn	NN 788368	51/52	E379	
870	Srón a' Chaoineidh	NN 771369	51/52	E379	
697	Creag an Sgliata	NN 769399	52	E48/E379	
665	Meall a' Choire Chreagach	NN 792412	52	E379	

	Height	Name	Grid Ref		
☐	888	△Creagan na Beinne	NN 744368	51/52	E48/E379
☐	716	Beinn Bhreac	NN 733402	51/52	E48
☐	759	Ciste Buide a' Chlaidheimh	NN 729351	51/52	E48
☐	682	Tullich Hill	NN 704367	51/52	E48
☐	833	Meall nan Oighreag	NN 704340	51/52	E48
☐	879	△Creag Uchdag	NN 708323	51/52	E48
☐	850	Creag nan Eun	NN 728318	51/52	E48
☐	619	Meall Mathaig	NN 708273	51/52	E47
☐	612	Meall nam Fiadh	NN 697275	51	E47
☐	615 est	Meall nam Fiadh Iar	NN 691272	51	E47
☐	712	Creag Ruadh	NN 677293	51	E47
☐	693 est	Meall Daimh	NN 664307	51	E47/E48
☐	682	Ruadh Mheall (Ruadh Bheul)	NN 677314	51	E48
☐	628	Meall Odhar	NN 639322	51	E48
☐	637	Creag Gharbh	NN 632327	51	E48
☐	620	Cul na Creige	NN 621298	51	E39/E48/E377
☐	612	Am Bacan	NN 607294	51	E48

30

Height	Name	NGR	OS L	OS E	Ascent
663	Meall na Cloiche	NN 590277	51	E46	
719	Meall Buidhe	NN 576276	51	E46	
705	Beinn Leabhainn	NN 575283	51	E46	
659	Srón Mhór	NN 647264	51	E47	
672	Creag Each	NN 652263	51	E47	
	Monzievaird:				
786	Cárn Chois	NN 792278	51/52	E47/E379	
931	▲Beinn Chonzie (Beinn a' Chonnaich)	NN 773308	51/52	E47/E379	
755	Biorach a' Mheannain	NN 783318	51/52	E379	
731 est	A' Chairidh	NN 795314	51/52	E379	
789	△Auchnafree Hill	NN 809308	52	E47/E379	
787	Cárn Choinneachain	NN 818288	52	E47/E379	
771	Stonefield Hill	NN 833297	52	E47/E379	
698	Meall Dubh	NN 863302	52	E47/E379	
648	Meall Tarsuinn	NN 877297	52/58	E47/E379	
619	Beinn Fendoch	NN 885305	52	E47/E379	

Logiealmond:
Meall Reamhar   NN 922332   52   E379
Meall nan Caorach   NN 929339   52   E379

☐ 620
☐ 624

Beinn Laoigh, Glen Fyne & Arrochar

One of the most distinctive mountains in Scotland is The Cobbler, the common name for Ben Arthur, which rises west of Arrochar. The rocky façade and pinnacle on the summit makes the hill a popular climb. The cliff face is also popular with rock-climbers. The Cobbler's prominence makes it eligible to be a Corbett, rising to 2,891 feet. Near to it is Beinn Narnain, a Munro, and further north rises Beinn Ime, a considerable mountain in excess of 1,000 metres. Sometimes these three hills are climbed in a single expedition. To the west of Beinn Ime is Beinn Luibhean, a second Corbett in the vicinity, rising high above the Rest and be Thankful pass and fairly quickly ascended from the car park there.

Located in the centre of this group of mountains is Beinn Mheadhoin, or Ben Vane, which just qualifies as a Munro, being 3,004 feet above sea level. Most ascents are from Inveruglas, a track forming part of the Cowal Way heading up the glen, followed by a path up the eastern ridge.

Also from Inveruglas an ascent can be made of another Munro, Ard Mhurlaig, or Ben Vorlich, which has two separate tops, the southern one being the higher. The northern top, although listed as a top by Munro, fails to qualify as a mountain for this list. An ascent over Mullach Coire nan Each leads on to the southern ridge. A circular route brings one back down over the two Little Hills, a complicated rocky landscape making the route slightly circuitous.

Rising above the north side of Glen Kinglas is Stob Coire Creagach and Binnein an Fhidhleir, the former a Corbett though the hill is better known by the latter's name, forming twin tops on the same ridge.

A remote Corbett is Meall an Fhúdair, which with Troisgeach form a substantial hill rising to the west of Glen Falloch. It rises to 2,508 feet, just qualifying to make Corbett's list.

North of Gleann nan Caorann and forming a ring of mountains around Glen Cononish, is a high range. Four of them are Munros and a fifth mountain qualifies as a Corbett. The eastmost Munro is Beinn Dubhchraig, usually ascended from Dalrigh. The ridge beyond Bealach Buidhe can be followed to the summit of Beinn Oss, a considerable mountain at 3,374 feet.

31

The highest mountain in this group is Beinn Laoigh, at 1,130 metres, or 3,708 feet. The summit cairn stands on a fairly flat top, which has two points vying for precedence, the southerly apparently winning, though climbers will probably visit both just in case. The northern top has a cairn, and between the two on the east side is the steep slopes and rocks at the head of Coire Gaothach, with the Central Gully being a notable winter climb for experienced mountaineers. It holds its snow long into the summer months. The mountain is part of the Beinn Laoigh National Nature Reserve, established to protect the saxifrages and lichens that grow there. The tiny lochan in Coire an Lochain is thought to be the source of the River Tay, Scotland's longest river.

West of Beinn Laoigh is a fourth Munro, Beinn a' Chleibh, which is only eight feet into Munro territory. It is located near to Glen Lochy, from where it requires a steep climb through the forests and up over the ridge of Stob Dubh, or else through Fionn Choirein to reach the top.

North-east of Beinn Laoigh is Beinn Chuirn, a substantial Corbett at 2,887 feet. Most ascents are made from Glen Cononish, passing the gold mine at the head of the glen.

To the south-west of Beinn Laoigh the hills are lower, gradually diminishing until Loch Fyne is reached. In amongst them is Beinn Bhuidhe, a Munro at 3,106 feet. A track from Glen Fyne ascends much of the southern slopes. Around it are a number of lower mountains, including Stob Choire Dhuibh and Ceann Garbh.

Height		Name	NGR	OS L	OS E	Ascent
656	☐	Meall Odhar	NN 297299	50	E39/E377	
880	☐	△ Beinn Chuirn	NN 280292	50	E39/E377	
773	☐	Beinn Chuirn Beag	NN 274284	50	E39/E377	
652	☐	Fiarach	NN 345261	50	E39/E377	
978	☐	▲ Beinn Dubhcraig	NN 307254	50	E39/E377	
1029	☐	▲ Beinn Oss	NN 287252	50	E39/E377	
1130	☐	▲ Beinn Laoigh (Ben Lui)	NN 266263	50	E39/E377	
916	☐	▲ Beinn a' Chleibh	NN 251255	50	E39/E377	
		Glen Fyne:				
743	☐	Meall nan Gabhar	NN 235241	50	E39	
739	☐	Meall nan Tighearn	NN 238234	50	E39	
636	☐	Beinn Bhalgairean	NN 203241	50	E39	
625	☐	Beinn Bhreac	NN 203216	50/56	E39	
694	☐	Beinn an t-Sithein	NN 182195	50/56	E360	
803	☐	Ceann Garbh	NN 222203	50/56	E39	
901	☐	Stob Choire Dhuibh	NN 213193	50/56	E39	

948	▲ Beinn Bhuidhe	NN 203187	50/56 E39
870	Stac a' Chuirn	NN 190182	50/56 E39
680	Beinn Chas	NN 199162	50/56 E39
733	Troisgeach	NN 290193	50/56 E39
764	△ Meall an Fhudair	NN 271192	50/56 E39
684	Beinn Damhain	NN 282172	50/56 E39
645	Maol Breac	NN 259158	50/56 E39
817	△ Stob Coire Creagach	NN 230109	50/56 E37/E39
811	Binnein an Fhidhleir	NN 214107	50/56 E37/E39

Arrochar:

647	Stob nan Coinnich Bhacain	NN 303146	50/56 E39
943	▲ Ben Vorlich (Ard Mhurlaig)	NN 295124	50/56 E39
808	Little Hill (West Top)	NN 303124	50/56 E39
793	Little Hill (East Top)	NN 308123	50/56 E39
782 est	Mullach Coire nan Each	NN 304110	50/56 E39
773	Beinn Dhubh	NN 274116	50/56 E39
916	▲ Beinn Mheadhoin (Ben Vane)	NN 277098	56 E39

31

Height	Name	NGR	OS L	OS E	Ascent
888	Beinn Chorranach	NN 254096	56	E39	
1011	▲ Beinn Ime	NN 255084	56	E39	
858	△ Beinn Luibhean	NN 243079	56	E37/E39	
858	Ceile nan Creasaiche (Arthur's Seat)	NN 260057	56	E39	
884	△ The Cobbler (An Greasaiche Crom)	NN 259058	56	E39	
867 est	Sgurr Tuath nan Greasaiche	NN 261060	56	E39	
926	▲ Beinn Narnain	NN 272067	56	E39	
781	Creag Tharsuinn	NN 278075	56	E39	
849	A' Chrois	NN 289077	56	E39	

Cowal

Cowal is that part of Argyll that is bounded by the sea lochs Long and Fyne, stretching south from the passes of Glen Kinglas and Glen Croe. Most of the slopes are afforested, leaving the summits rising above the tree line. Part of Loch Lomond and the Trossachs National Park extends into Cowal, encompassing most of the mountains within this list.

At the north-east end of Cowal is the Ardgoil Estate, a peninsula between Lochs Long and Goil. This is Forestry Commission country, and the forest roads are useful for making access routes to the bases of the mountains. Cnoc Coinnich, The Brack and Beinn Donich are Corbetts, indicating that they are substantial hills in their own right. The lower Beinn Reithe is the most remote of the Ardgoil summits, as well as the lowest.

North of Gleann Mór are three summits, the eastmost, Beinn an Lochain, being Cowal's fourth Corbett. This mountain was in fact listed as a Munro for a period, it being thought that the summit just reached 3,000 feet. Munro's original tables included it, claiming the height as 3,021 feet, but in the 1970s it was measured again and found to be just 2,957 feet in height.

West of Lochgoilhead are five summits forming the western extremity of the national park. Again, their lower slopes are afforested by the Forestry Commission, but their upper slopes are enlivened by rock outcrops and small cliffs. The southernmost of these summits is Beinn Lochain, quickly ascended from Curra Lochain, reached by following the Cowal Way, a long-distance path.

On the opposite side of Bealach an Lochain rises Beinn Bheula, the fifth and final Cowal Corbett. This mountain has a number of smaller tops around it, the highest point being Caisteal Dubh. Flanking it are the tops of Cárnach Mór and Beinn Dubhain, the three summits achievable in the route round Coire Aodainn from Invernoaden to the west. At Túr nan Calman walkers can explore the caves located near to the summit of Cárnach Mór.

Between Loch Goil and Loch Eck is a ridge with five more summits on it. Sgurr a' Choinnich is the northernmost, followed by Beinn Bhreac and Cruach a' Bhuic, the former summit just meriting inclusion. Cruach Eighreach also just qualifies as a separate mountain, there being just 100 feet of re-ascent from Creachan Mór, its parent top.

Beinn Ruadh lies in solitude as far as summits in excess of 2,000 feet are concerned, between lochs Eck and Long. Lying within Forestry Commission property, it is usually climbed from Inverchapel to the west, where the forest opens up, or else from around Stronvochlan in Glen Finart to the east.

On the west side of Loch Eck, the quiet side, are three tops. Beinn Bheag being rarely ascended due to its remoteness and the need to walk half-way along Loch Eck to reach its foot. Its larger neighbour to the south, Beinn Mhór, is also quite remote, but is more readily ascended from Glenmassan, where a forestry road climbs the first stretch up the side of Allt Coire Mheasan. Being so high in this part of Cowal returns some extensive views on a clear day. A flank of Beinn Mhór rises in a conical peak overlooking the south end of Loch Eck. Clach Bheinn is reached from near to Benmore Botanic Garden, a break in the forest beyond Allt Corrach giving a clear route to the summit, and to Beinn Mhór beyond if desired.

The final two hills are random summits that rise in excess of 2,000 feet. Creag Tharsuinn is to the west of Garrachra Glen, part of a ridge of hills that can be climbed from a variety of starting points, including Strath nan Lub to the west, or Glen Massan to the east.

Cruach nan Capull is the southernmost peak in Cowal that rises above 2,000 feet. It makes this only just, by five feet, but it affords fine views down Loch Striven towards Bute. It is quickly ascended from either Corrachaive to the north or Invervegain to the south.

32

Height	Name	NGR	OS L	OS E	Ascent
	Ardgoil Estate:				
655	Beinn Reithe	NS 229986	56	E37	
764	△ Cnoc Coinnich	NN 233007	56	E37	
787	△ The Brack	NN 245031	56	E37/E39	
847	△ Beinn Donich	NN 219043	56	E37/E39	
901	△ Beinn an Lochain	NN 218079	56	E37/E39	
719	Beinn an t-Seilidh	NN 201080	56	E37/E39	
732	Stob an Eas	NN 186074	56	E37/E39	
611	Cruach nam Mult	NN 169056	56	E37	
639	Mullach Coire a' Chuir	NN 171034	56	E37	
658	Stob na Boine Druim-fhinn	NN 169025	56	E37	
619	Beinn Tharsuinn	NN 164017	56	E37	
703	Beinn Lochain	NN 160006	56	E37	
634	Cárnach Mór	NS 141992	56	E37	
779	△ Beinn Bheula	NS 154983	56	E37	
649	Beinn Dubhain	NS 143973	56	E37	

32

Height	Name	NGR	OS L	OS E	Ascent
661	Sgurr a' Choinnich	NS 159957	56	E37	
623	Beinn Bhreac	NS 163940	56	E37	
635	Cruach a' Bhuic	NS 169935	56	E37	
649	Cruach Eighrach	NS 181921	56	E37	
657	Creachan Mór	NS 186916	56	E37	
664	Beinn Ruadh	NS 156884	56	E37	
618	Beinn Bheag	NS 125932	56	E37	
741	Beinn Mhór	NS 108908	56	E37	
643	Clach Bheinn	NS 126886	56	E37	
643	Creag Tharsuinn	NS 087914	56	E37/E362	
612	Cruach nan Capull	NS 096796	63	E37/E362	

Luss Hills

The Luss Hills are located between Loch Lomond (to the east) and Loch Long (to the west). All are included within the Loch Lomond and the Trossachs National Park, though four of the western summits form the park's western boundary hereabouts. The northernmost summit, to the south of the Arrochar-Tarbet isthmus, is Beinn Riabhaich, or Ben Reoch, which rises steadily from the loch sides. Another three summits are in this stretch of upland countryside, north of Glen Douglas, with Beinn Bhreac being the tallest.

Glen Douglas, at its western end, is occupied by Ministry of Defence property, north-west of Doune Hill. However, access to this summit, which is the highest in this group, is readily reached from the glen, usually from Doune farm. Another route is over Beinn Eich from Glen Luss. Military restrictions also apply to the countryside west of Beinn a' Mhanaich, but the long ridge of The Strone to the summit is open from the Glenfruin Road near to Strone House.

Beinn Chaorach is the second tallest of the Luss Hills, usually climbed from Auchengaich in Glen Fruin. Beinn Tharsuinn and Balcnoc can also be included in a round of the Allt Baile a' Mhuilinn glen. Alternatively, the three summits can be done with Creag an Leinibh as a longer circular route from Glen Luss, starting at Luss itself.

North of Luss rises Beinn Dubh, the long ridge forming a steady climb from the village, Loch Lomond's water surface being almost at sea level. Views of the loch from the ridge are astounding, the islands in the wider southern part of the loch becoming more obvious the higher one climbs. Part of the same bulk is the taller summit of Mid Hill.

Most of this section falls within Luss Estate, the ancient property of the Colquhoun family, whose ancestral seat is at Rossdhu. The area west of the National Park boundary is part of Defence Estates.

33

Height	Name	NGR	OS L	OS E	Ascent
661	Beinn Riabhaich (Ben Reoch)	NN 308021	56	39	
632	Croit a' Chladaich	NN 313018	56	39	
681	Beinn Bhreac	NN 322000	56	39	
632	Tullich Hill	NN 294006	56	39	
642	Beinn Dubh	NS 335953	56	39	
657	Mid Hill (Cnoc Meadhonach)	NS 322963	56	39	
701	Cnoc an Duin Ear	NS 297974	56	39	
734	Doune Hill (Cnoc an Duin)	NS 261972	56	39	
703	Beinn Eich	NS 302947	56	38/39	
684	Cruach an t-Sithein	NS 275964	56	39	
709	Beinn a' Mhanaich	NS 269946	56	38/39	
713	Beinn Chaorach	NS 287924	56	38	
656	Beinn Tharsuinn	NS 291916	56	38	
693	Balcnoc (Stob Coire Fuar)	NS 302914	56	38	
658	Creag an Leinibh	NS 311919	56	38	

Balquhidder & Strath Gartney

Forming the heart of the Loch Lomond and Trossachs National Park, this section is bounded by Loch Lomond to the west, Glen Dochart to the north and Glen Ogle and Strathyre to the east. The mountains drop in height towards the south, apart from the notable Ben Lomond.

At the north-east corner, above Lochearnhead and Balquhidder, are the two Corbetts of Creag Mac Ránaich and Meall an t-Seallaidh. Other summits surround these hills, and though they make for attractive walking, they tend to be quiet.

The Braes of Balquhidder rise north of Loch Voil, rising in height to the west. The Stob, or Meall na Frean is a fairly flat summit hereabouts, overlooked by the taller and more attractive mountains west of the Monachyle Glen. Of these, Meall na Díge is a Munro top, and Stob Creagach is an attractive 900 mertre summit.

West again, rising steeply to the south of Glen Dochart, is Beinn Mór, or Ben More, even its name meaning large mountain. At 3,843, it is one of the taller mountains in the district, often appearing as a cone when viewed from afar. Immediately south of Beinn Mór is Stob Binnein, a second Munro that is almost as tall as its neighbour, failing to reach its height by 22 feet. It, too, is conical in appearance, and to its south is a Munro top, Stob Coire an Lochain, which, however, fails to qualify for this list.

To the west of the Benmore-Inverlochlarig glens rise a second pair of Munros, Cruach Ardrain and Beinn Tulaichean. The former is 3,428 feet in height, a prominent summit, usually climbed from Crianlarich to the north. Alternatively, this mountain can be climbed from Inverlochlarig to the south, including an ascent of Beinn Tulaichean. Stob Garbh to the north-east of Cruach Ardrain is a Munro top.

Westwards again is another block of mountains, with An Caisteal being the highest, a Munro of 3,265 feet. To its south is Beinn a' Chroin, the western top being the taller, whereas historically the eastern was regarded as being superior. Both qualify as mountains for this guide. Westwards again is Beinn Chabhair, a Munro of 3,053 feet.

A range of mountains occupies the landscape between lochs Voil and Katrine, though none reaches Munro status. The nearest to do so are Stob a' Choin, a Corbett that rises near to the head of the Balquhidder valley, Beinn Bhán and Beinn Ledi, two more Corbetts that are located further east. Beinn Ledi is a notable peak, visible from across Stirling and southern Perth shires, rising steeply above the Pass of Leny. Slightly lower, and rising south of Loch Doine, is Beinn Stacach, a fourth Corbett.

Another Corbett can be found on the east side of Loch Lomond, north of Loch Arklet. Beinn a' Choin presents itself as a rocky knoll, the south-western slopes being part of Inversnaid Nature Reserve.

The hills immediately south of Loch Arklet do not reach mountain height until the rounded Cruinn a' Bheinn is reached. From here the northern flanks and corries of Ben Lomond are seen, Scotland's most southerly Munro and a considerable mountain visible from many miles away to the south and east. This mountain is owned by the National Trust for Scotland, and forms Ben Lomond National Memorial Park. A couple of popular paths make their way to the summit from Rowardennan, on Loch Lomond-side.

South of Loch Katrine, in The Trossachs proper, rises Ben Venue and a few other summits. Ben Venue, or Bheinn Uamh to give it its proper name, is a popular climb from the Pass of Trossachs. It was referred to in Sir Walter Scott's poetry.

Various parts of this section are in public ownership, allowing free access. In addition to Ben Lomond, much of the land is owned by the Forestry Commission, Glen Finglas is owned by the Woodland Trust, its largest reserve, Inversnaid by the RSPB and Loch Katrine by Scottish Water.

34

The Mountains of Great Britain

Height	Name	NGR	OS L	OS E	Ascent
670 est	Meall Reamhar	NN 570247	51	E46	
809	△Creag Mac Ranaich	NN 545256	51	E46	
852	△Meall an t-Seallaidh	NN 542234	51	E46	
812	Cam Chreag	NN 539241	51	E46	
817	Meall an Fhiodhain	NN 532244	51	E46	
789	Leum an Eireannaich	NN 521249	51	E46	
675	Meall Reamhar	NN 513227	51	E46	
734	Stob Caol	NN 492221	51	E46	
753	Meall na Frean - The Stob	NN 492231	51	E46	
906	Stob Creagach	NN 459232	51	E46	
966	Meall na Dige	NN 452226	51	E46	
1165	▲Stob Binnein	NN 434226	51	E46	
1174	▲Beinn Mhór (Ben More)	NN 433244	51	E46	
857	Stob Coire Bhuidhe	NN 409229	51	E39	
959	Stob Garbh	NN 410221	51	E39	
1046	▲Cruach Ardrain	NN 407211	51/56	E39	

34

Height	Name	NGR	OS L	OS E	Ascent
814	Meall Dhamh	NN 398217	56	E39	
946	▲ Beinn Tulaichean	NN 416196	56	E46	
830	Stob Glas	NN 404202	56	E39	
940	Beinn Ear a' Chroin	NN 395186	50/56	E39	
942	▲ Beinn a' Chroin	NN 385185	50/56	E39	
995	▲ An Caisteal	NN 379193	50/56	E39	
654	Beinn Glas	NN 344190	50/56	E39	
715	Stob Creag an Fhithich	NN 349191	50/56	E39	
691 est	Càrn a' Gharbh Bhealaich	NN 353185	50/56	E39	
719	Meall nan Tarmachan	NN 358185	50/56	E39	
933	▲ Beinn Chabhair	NN 368179	50/56	E39	
666	Parlan Hill	NN 353170	50/56	E39	
747	Meall Mór	NN 384151	50/56	E39	
727	Stob an Duibhe	NN 397154	50/56	E39	
714	An Garadh	NN 405142	56	E39	
869	△ Stob a' Choin	NN 417161	56	E46	

	Height	Name	Grid Ref		Map
☐	836	Meall Reamhar	NN 425156	56	E46
☐	686	Stob Breac	NN 447165	57	E46
☐	669	An Stuchd	NN 447148	57	E46
		Western Strathyre Forest:			
☐	715	Taobh na Coille	NN 466152	57	E46
☐	771	△Beinn Stacach (Stob Fear-Tomhais)	NN 474163	57	E46
☐	697	Creagan nan Sgiath	NN 486143	57	E46
☐	658	Creag Mhór	NN 511185	57	E46
☐	674	Meall Cala	NN 508128	57	E46
☐	821	△Beinn Bhán (Ben Vane)	NN 535137	57	E46
☐	715	Ardnandave Hill	NN 567125	57	E46
☐	722	Bioran na Circe	NN 558118	57	E46
☐	879	△Beinn Ledi	NN 562098	57	E46
☐	638	Stúc Odhar	NN 551088	57	E46
		Aberfoyle Hills:			
☐	727	Bheinn Uamh Ear	NN 477061	57	E46
☐	729	Bheinn Uamh (Ben Venue)	NN 474063	57	E46

34

34

Height	Name	NGR	OS L	OS E	Ascent
684	Stob an Lochain	NN 466050	57	E46	
700	Beinn Bhreac	NN 458059	57	E46	
703	Beinn Chochan	NN 453057	57	E46	
	Buchanan:				
613	Stob nan Eighrach	NN 342144	50/56	E39	
651 est	Maol an Fhithich	NN 349139		E39	
770	△Beinn a' Choin	NN 354130	50/56	E39	
694	Maol Mór	NN 373120	50/56	E39	
655	Stob an Fhainne	NN 359111	50/56	E39	
632	Cruinn a' Bheinn	NN 365052	56	E39	
974	▲Ben Lomond (Beinn Laomuinn)	NN 367029	56	E39	
778	Cárn a' Bhealach Buidhe	NN 361028	56	E39	

Glen Artney

The Forest of Glenartney is an extensive sporting estate extending over most of Glen Artney. The glen is surrounded by tall hills, some of which reach above 2,000 feet. To the west of the deer forest, falling within the Loch Lomond and Trossachs National Park, is the Glen Ample area, much of which is afforested as part of Strathyre Forest. A major part of this section is part of Drummond Castle Estate, stretching from Muthill in the east as far west as Beinn Each. Strathyre is Forestry Commission countryside, whereas some of the southern stretches are operated as large farms, or else part of Moray Estates.

To the south of the road from Comrie to Lochearnhead is Mór Bheinn and Ben Halton, two rocky summits just in excess of 2,000 feet. West of Ben Halton are three mountain tops, of which Creag na h-Eararuidh is just slightly taller than the more distinctive Beinn Dearg. The third summit, Srón na Maoile is the least distinctive.

Arranged along the southern side of Loch Earn is a line of summits, many of which are over 2,000 feet but which fail to qualify as a mountain due to the re-ascent rule. Amongst the many tops, those that do count are Beinn Fuath at the eastern end, south of St Fillans, Cárn a' Bhothain, Meall nan Saighdearan, Meall Reamhar, Beinn Bhán and Creagan an Lochain.

South of this ridge can be found the tall summit of Meall na Fearna, a Corbett often climbed from either Glen Artney to the south or Glen Vorlich to the north. On the west side of Srath a' Ghlinne are three tops that form a ridge from Meall na Fearna—Stob Chalum Mhic Griogair, Srón Bhuidhe and Stúc Gharbh. These three tops can be climbed in a round to Meall na Fearna from Glen Artney.

South-east of Lochearnhead rise two Munros, usually ascended in a single expedition. The more readily ascended is the taller of the two, Ben Vorlich, a path from near Ardvorlich making its way up a northern ridge. A path continues on over the Bealach an Dubh Choirean to Stúc a' Chroin, which is almost as tall as Ben Vorlich. From the north, Stúc a' Chroin appears as a pointed summit, rising above Coire Fhuadaraich. A return to Ardvorlich can be made over Creagan nan Gabhar and Beinn Odhar. Stúc a' Chroin is the target in an annual hill race in May, with 5,000 feet of ascent from Callander and back possible in around two hours. On the northern slopes of the hills is a cairn with a memorial to Donald Stuart, founder of the Falkirk Mountaineering Club. These two Munros are visible for miles around, and can be seen from as far away as Fife.

West of Glen Ample is a solitary mountain, Sgiath a' Chaise, the woods of Strathyre Forest almost reaching the top. East of this top, and to the south of Stúc a' Chroin is a range of summits, the tallest being Beinn Each. Its eastern slopes are steep, adding to its appeal. This summit qualifies as a Corbett, being 2,667 feet in height. Ascents to the top are often made from Ardchullarie Mór, on the side of Loch Lubnaig.

The final group of mountains in this section form the upper reaches of the Braes of Doune. There are four summits in excess of 2,000 feet, Beinn nan Eun at the eastern end, its rocky slopes dropping to Finndhu Glen. To its west, more rounded and less rocky, is Beinn Odhar, its summit marked by a cairn. On the southern slopes of this top is the Braes of Moray wind farm, the regular line of turbines ranged across the slope destroying the quiet rural nature that this area once enjoyed.

Uamh Bheag rises to the west of the wind farm, the tallest summit in this small range. The summit is marked by a cairn at the junction of fences, to the west of the eastern top which is marked by a trig point. A shoulder of this hill rises sufficiently to qualify as a mountain in its own right—Meall Clachach. Its prominence is probably just 100 feet and no more. Uamh Bheag and Uamh Mhór to the south of it are named after a few small caves that exist on the latter hill's slopes.

The Mountains of Great Britain

Height	Name	NGR	OS L	OS E	Ascent
	Forest of Glenartney:				
640	Mór Bheinn (Morven)	NN 716212	51/52/57	E47	
621	Ben Halton	NN 720203	51/52/57	E47	
707	Beinn Dearg	NN 697198	57	E47	
708	Creag na h-Eararuidh	NN 685190	57	E47	
618	Srón na Maoile	NN 690176	57	E47	
661	Beinn Fuath	NN 690217	57	E47	
642	Càrn a' Bhothain	NN 682216	57	E47	
681	Meall nan Saighdearan	NN 672212	51/57	E47	
662	Meall Reamhar	NN 664211	51/57	E47	
692	Beinn Bhan (Black Craig)	NN 651210	51/57	E47	
685	Creagan an Lochain	NN 646209	51/57	E46/E47	
739	Beinn Domhnuill	NN 644199	57	E46/E47	
717	Càrn na Bhealaich Gliogarsnaich	NN 643191	57	E46/E47	
809	△ Meall na Fearna	NN 651187	57	E47	
741	Stob Chalum Mhic Griogair	NN 660190	57	E47	

35

35

Height	Name	NGR	OS L	OS E	Ascent
701	Srón Bhuidhe	NN 662184	57	E47	
636	Stúc Gharbh	NN 668174	57	E47	
985	▲Ben Vorlich (Beinn Mhùrlaig)	NN 629189	57	E46/E47	
727 est	Creagan nan Gabhar	NN 615200	51/57	E46	
741 est	Beinn Odhar (Ben Our)	NN 616206	51/57	E46/E47	
975	▲Stúc a' Chroin	NN 617174	57	E46/E47	
646	Meall Odhar	NN 645145	57	E46/E47	
	Braes of Doune:				
621	Meall Clachach	NN 688126	57	E47	
666	Uamh Bheag	NN 691119	57	E47	
626	Beinn Odhar	NN 714128	57	E47	
631	Beinn nan Eun	NN 722131	57	E47	
	Easter Strathyre Forest:				
694	Mullach Mhòr	NN 617146	57	E46/E47	
765	Meall na Caora	NN 608151	57	E46	
813	△Beinn Each	NN 601158	57	E46	

☐ 735	Cárn a' Bhealaich Ghlais	NN 603169	57	E46
☐ 645	Sgiath a' Chaise	NN 583169	57	E46

35

Ochil Hills

The Ochil Hills form a high stretch of countryside separating Perthshire from Clackmannanshire. From near to Bridge of Allan, just north of Stirling, they stretch in a generally eastward direction, gradually lowering in height until around Bridge of Earn. The summits that are over 2,000 are all located in the western section, south-west of Glen Devon and rising steeply up from the 'Hillfoot' villages—Alva, Tillicoultry and Dollar.

The westmost summit is Blairdenon Hill, the summit of which is located at the junction of three regions—Stirling, Clackmannanshire and Perth & Kinross. In the pass to the west, before Greenforet Hill, is a memorial to A. J. Cuthbertson, a pilot killed in an air crash here in 1957.

Ben Cleuch is the tallest of the Ochils, and two other summits are located on its shoulders—Ben Ever and Andrew Gannel Hill. Most ascents of these hills are made from the south, Ben Cleuch being ascended steeply from Mill Glen, at Tillicoultry. This takes one over The Law, a subsidiary top. The circuit of the steep Mill Glen can be made over Ben Ever, dropping back to the start, or else to Alva. On the summit is a viewpoint indicator.

King's Seat Hill rises steeply above Tillicoultry and Dollar, paths from both communities making their way to the summit. Views of the Forth plain as far as Edinburgh and the Pentland Hills are made from here.

The three remaining Ochil mountains are not so well known, being less significant in their connection to the hillfoot towns, and more remote. The three form a single ridge, Tarmangie Hill to the west, with Whitewisp Hill being a second summit on the same bulk, the re-ascent to Whitewisp Hill being just 32 metres. A longer gap separates Innerdownie, a summit that just makes the 2,000 foot contour. It is surrounded by the Glendevon Forest, parts of which are owned by the Woodland Trust Scotland.

Height	Name	NGR	OS L	OS E	Ascent
611	Innerdownie	NN 966031	58	E366	
643	Whitewisp Hill	NN 955013	58	E366	
645	Tarmangie Hill	NN 942014	58	E366	
648	King's Seat Hill	NN 933000	58	E366	
670	Andrew Gannel Hill	NN 919005	58	E366	
721	Ben Cleuch	NN 902006	58	E366	
622	Ben Ever	NN 893001	58	E366	
631	Blairdenon Hill	NN 865018	58	E366	

36

223

Arran

Although Arran is often described as being 'Scotland in miniature', the mountains on it are full-blown examples. Being an island, most of the ascents start from near sea-level, meaning that the climbs are considerable, and the rugged nature of the mountains make them difficult to traverse in some areas. Places such as the ridge known as A' Chir (the comb) and Ceum na Caillich (the witch's step) require some degree of scrambling, and can be extremely dangerous in winter or rough weather. The great peak of Cir Mhór has a small pointed summit, some sides of which are difficult to ascend. Much of northern Arran comprises granite which has forced its way into Dalriadian schists.

The most famous summit on Arran, as well as the highest, is Goatfell, which rises above Brodick Bay, from where its cone is a distinctive landmark. Most ascents are made by following the path from near Brodick Castle. Goatfell is a Corbett and at 2,867 feet is not too far away from Munro height. The summit contains a direction indicator, pointing out the local mountains as well as some distant ones. On a clear day it is possible to see Ireland, the Galloway Highlands, Cruachan and Ben Lomond. Goatfell earned a place in history in 1889 when Edwin Rose was murdered on it by John Watson Laurie.

The Stacach ridge stretching north from Goatfell leads to other summits, including Mullach Buidhe and North Goatfell, both prominent enough to qualify as mountains. This ridge terminates in the pointed Cioch na h-Oighe, a summit worth climbing in its own right, even although it has insufficient re-ascent to qualify as a distinct mountain.

A deep descent to the bealach at the head of glens Rosa and Sannox allows a further ascent of Cir Mhór, a steep scramble required to reach the summit of this Corbett. Like many of the mountains hereabouts, the rock is igneous. Cir Mhór is one of the

37

most attractive of mountains on Arran, if not Scotland, its rugged pinnacle surrounded by steep cliffs.

Suidhe Fhearghas and Ceum na Caillich can be climbed en route to Caisteal Abhail, another prominent Corbett summit. Ceum na Caillich is a pinnacle high above a narrow ridge, sufficiently prominent to qualify as a mountain in its own right. Caisteal Abhail is Arran's second highest peak, at 2,818 feet.

The third Corbett on the island is Beinn Tarsuinn, a great hulk of a summit on the west side of Glen Rosa, one of the island's most attractive glens. Reaching it from Cir Mhór allows climbers to traverse the interesting A' Chir, a rocky ridge which requires some scrambling skill and which merits listing as a mountain. Beinn Tarsuinn is itself a rocky ridge-like top, a long spur heading south. This ridge is followed when ascending Beinn Nuis, the summit cairn perched on the edge of the steep Creag na h-Iolaire. The final mountain in the eastern Arran group of mountains is Beinn a' Chliabhain. It rises to the west of Glen Rosa, the longish ridge being fairly tame compared with the other Arran peaks.

On the west side of the island, above Pirnmill, is a range of three mountains. The southernmost is Beinn Bharrain, the cairn at Caisteal na h-Iolaire marking the western summit. Slightly higher is the rounded top of Mullach Buidhe, regarded as being part of Beinn Bharrain. A curved ridge leads north to the third mountain, Beinn Bhreac,. To its north is Meall Donn, which almost qualifies as a mountain, but whose re-ascent is just too little. It overlooks the attractive Coire Fhionn Lochan.

The land to the east side of Beinn Nuis, Beinn Tarsuinn and A' Chir, south of Cir Mhór and Mullach Buidhe, including Goatfell, belongs to the National Trust for Scotland, being part of its Brodick Castle Estate. This was given to the trust in 1958 by Lady Jean Fforde. Glen Sannox and the north side of Caisteal Abhail form part of Sannox Estate, whereas the western hills, and the land from there eastwards to the Sannox and Brodick estates, form Dougarie Estate.

The Mountains of Great Britain

Ascent										

Height	Name	NGR	OS L	OS E
660 est	Suidhe Fhearghas	NR 986451	69	E361
727	Ceum na Caillich	NR 977443	69	E361
859	△Caisteal Abhail	NR 969443	69	E361
799	△Cir Mhór	NR 973431	69	E361
745	A' Chir	NR 966421	69	E361
826	△Beinn Tarsuinn	NR 959412	69	E361
792	Beinn Nuis	NR 955399	69	E361
675	Beinn a' Chliabhain	NR 970407	69	E361
829	Mullach Buidhe	NR 99343427	69	E361
818	North Goatfell	NR 990423	69	E361
874	△Goatfell	NR 991416	69	E361
717	Beinn Bharrain	NR 896422	69	E361
721	Mullach Buidhe	NR 902427	69	E361
711	Beinn Bhreac	NR 907442	69	E361

Galloway

The high hills of the Galloway Highlands rise in south west Scotland. Most of the area is owned by Scottish Land and Forestry, meaning that their lower slopes are afforested, but there is still considerable areas of wild countryside to explore.

The most southerly mountains are the three summits that form the Cairnsmore of Fleet massif. This rises east of Newton Stewart in a great granite mass. Most ascents are made from Cairnsmore House, a path making its way to the summit. There can be found a prehistoric burial cairn and a memorial to aeroplane crashes that appear to have been common in this mountain. Knee of Cairnsmore to the south has a second Bronze Age cairn on the summit, and a third top, Meikle Mulltaggart, rises to the north-east of the main summit.

The Minnigaff Hills rise to the north of Newton Stewart. Larg Hill and Lamachan Hill appear as two great lumps when viewed from the south, with steep corries on their slopes. Curleywee is more rugged, and appears to have more of a pointed summit. The three hills are often climbed in a single expedition, perhaps starting at Glen Trool to the north.

In Glen Trool is a large boulder commemorating Robert the Bruce's victory over the English in 1307. He used these remote hills to hide within, and a number of places have connections with him. The main path to the top of the south of Scotland's highest mountain, Merrick, starts from the boulder, climbing up by the side of the Buchan Burn and its waterfalls to the summit of Ben Yellary, and from there to the Merrick's top itself. This mountain qualifies as a Corbett and has some steep corries on its northern flanks. It is said that the furthest view in Britain can be made from the summit, as far as Snowdon in Wales.

Stretching to the north is a range of mountains,

known colloquially as the Awful Hand. The mountains within it are Kirriereoch Hill, on the Ayrshire border, Tarfessock, Shalloch on Minnoch and Caerloch Dubh. Shalloch on Minnoch is a Corbett. Walking the length of the full ridge is an energetic climb.

Three more mountains are arranged in a north-south range to the east of the Merrick. These summits are granite, and at regular intervals the rock breaks through the surface to form cliffs and boulder fields. Mullwharchar is the furthest north, at one time a proposed location of nuclear waste dumping. South of it is the rocky summit of Dungeon Hill, which, though lower than many summits hereabouts, is one of the more interesting, especially when seen from across the Silver Flow. The third granite summit is Craignaw, which rises in the midst of a sea of lochans.

The final mountain range in Galloway is the Rhinns of Kells, and its northern extension, the Carsphairn range. Meikle Millyea is the furthest south, a not inconsiderable hill. Milldown follows, then it is the great summit of Corserine, a Corbett. To the north of this is the narrow ridge that peaks at Carlin's Cairn, another spot with associations with Robert the Bruce.

The ridge continues north over Meaul and Bow to Coran of Portmark. On Meaul is a memorial to a Covenanter who was killed there. On the col to the east is the King's Well and King's Stone, where Bruce also rested, and the outlying top of Cairnsgarroch.

Almost every summit listed falls within the Galloway Forest Park, or else partially does. Only the Fleet hills do not. East of Cairnsmore of Fleet is land forming part of a national nature reserve. The Forrest estate and the Garryhorn lands are not in public ownership. Many of the lochs on the lower edges of the hills have been dammed to provide water for hydro-electric generation, such as Loch Doon to the north, where the castle was rebuilt from its island, and Clatteringshaws Loch to the south, which is a totally man-made loch. Loch Trool is natural, as are the more remote Loch Dee, Loch Valley, Loch Neldricken, Loch Enoch and Loch Macaterick.

The Mountains of Great Britain

Height	Name	NGR	OS L	OS E	Ascent
	Fleet:				
☐ 656	Knee of Cairsmore	NX 509654	83	E319	
☐ 711	Cairnsmore of Fleet	NX 502671	83	E319	
☐ 612	Meikle Mulltaggart (Meall an t-Sagairt Mór)	NX 512678	83	E319	
	Minnigaff Hills:				
☐ 674	Curleywee	NX 454764	77	E319	
☐ 717	Lamachan Hill	NX 435770	77	E319	
☐ 676	Larg Hill	NX 424757	77	E319	
☐ 656	Millfore (Meall Fuar)	NX 478754	77	E319	
	Awful Hand:				
☐ 719	Ben Yellary (Beinn na h-Iolaire)	NX 414839	77	E318	
☐ 843	△ Merrick (Meurach)	NX 427855	77	E318	
☐ 786	Kirriereoch Hill (Cárn a' Choire-ribhaich)	NX 421870	77	E318	
☐ 697	Tarfessock (Torr a' Fasach)	NX 409892	77	E318	
☐ 775	△ Shalloch on Minnoch	NX 408906	77	E318	
☐ 659	Caerloch Dhu	NX 400920	77	E318	

38

Height	Name	NGR	OS L	OS E	Ascent
☐ 645	Craignaw	NX 459833	77	E318	
☐ 616 est	Dungeon Hill	NX 461861	77	E318	
☐ 692	Mullwharchar (Meall Adharc-fhiadhach)	NX 453867	77	E318	
	Rhinns of Kells:				
☐ 748 est	Meikle Millyea (Meall Liath Mor)	NX 516826	77	E318	
☐ 738	Milldown (Meall Donn)	NX 511839	77	E318	
☐ △ 814	Corserine	NX 498870	77	E318	
☐ 807	Carlin's Cairn (Cárn na Cailleach)	NX 497883	77	E318	
☐ 695	Meaul (Meall)	NX 500909	77	E318	
☐ 659	Cairnsgarroch (Cárn Sgarach)	NX 515913	77	E318	
☐ 613	Bow	NX 508928	77	E318	
☐ 623	Coran of Portmark	NX 509936	77	E318	

38

Kyle & Carsphairn

Stretching from upper Nithsdale to the Glenkens is a high expanse of hill country, much afforested on the lower slopes and now blighted in many areas with wind turbines and associated tracks. At the northern end are the Afton hills, arranged along the eastern side of Glen Afton, the valley with the stream famed in Robert Burns' work, *Flow Gently Sweet Afton*. The northernmost summit in the Afton range is also the highest, Blackcraig Hill, which rises steeply above the glen, its west side having a rocky face.

Extending south in a sweep around the headwaters of the Afton Reservoir are four other summits—Blacklorg Hill, Meikledodd Hill, Alwhat and Alhang. These are arranged along the Ayrshire-Dumfriesshire/Kirkcudbrightshire boundary, though Meikledodd Hill's true summit is in Kirkcudbrightshire.

On the west side of the headwaters of the Afton rises Windy Standard, the summit of which has turbines arranged across it. So too has the lower height of Dugland, something of an issue regarding listing. Historically it was indicated on maps as having a 2,000 feet contour line, but later 1:25,000 maps indicated its height as just 608 metres. Current 1:50,000 maps now show its height as 612 metres, meaning that it should be listed as a mountain once more.

Three more mountains rises to the south. Moorbrock Hill is located furthest east, an old mineral prospecting track leading to near the summit cairn. It has a steep eastern slope, hereabouts known as a gairy, as do the western two hills—Ben Inner and Cairnsmore of Carsphairn. The latter hill is the tallest in this section, reaching 797 metres, or 2,614 feet, making it the district's only Corbett. This mountain appears as a great whaleback when viewed from the south, from where most ascents commence. These usually start at the Green Well of Scotland, just north of Carsphairn.

39

Height	Name	NGR	OS L	OS E	Ascent
	Kyle:				
642	Alhang	NS 642010	77	E328	
628	Alwhat	NS 647020	77	E328	
643	Meikledodd Hill	NS 660027	77	E328	
681	Blacklorg Hill	NS 654042	77	E328	
700	Blackcraig Hill	NS 647064	77	E328	
	Carsphairn Forest:				
612	Dugland	NS 602009	77	E328	
698	Windy Standard	NS 620015	77	E328	
650	Moorbrock Hill	NX 621984	77	E328	
710	Ben Inner (Beinn an Innear)	NX 605972	77	E328	
797	△Cairnsmore of Carsphairn (Cárn Mór)	NX 594980	77	E328	

Lowther Hills

The Lowther Hills form one or more vertebrae in the spine of the Southern Uplands. They occupy the land between Annandale and Nithsdale, forming the ring of hills around the headwaters of the River Clyde. The A702 road divides the main summits into two groups, the northern hills (often referred to as the Lead Hills) and the southern group.

The northern range are dominated by the two highest hills, Green Lowther and Lowther Hill, the summits of which are covered with wireless and other transmission masts. A road from Scotland's highest village, Wanlockhead, can be followed all the way to the top. The Southern Upland Way also traverses Lowther hill, one of the highest points on the route.

North east of Green Lowther are a series of lower heights, and though rounded hills, they have steep sides to them. These are Dun Law and Lousie Wood Law. West of Lowther Hill is the strangely named East Mount Lowther, its top marked by a view indicator. The Southern Upland Way also crosses Cold Moss, a rather flat summit.

The southern hills are similar, if wilder, than the Lead Hills. Comb Law is the furthest north of these, and due south of it is Rodger Law. Continuing south-west is Ballencleuch Law and Scaw'd Law, the latter on the county boundary. So too are the next three hills—Wedder Law, Gana Hill and Earncraig Hill. Earncraig Hill only just qualifies as a mountain.

The final summit in the Lowther Hills, and the furthest south, is Queensberry. A fairly rounded height, it is visible for miles around, being one of Dumfriesshire's better known hills. The name was also used as a title by the Duke of Queensberry, now absorbed in the Duke of Buccleuch and Queensberry title. Buccleuch Estates own most of the Dumfriesshire parts of this section, Dun Law and Lousie Wood Law being in the Marquis of Linlithgow's Leadhills Estate.

40

The Mountains of Great Britain

Height	Name	NGR	OS L	OS E	Ascent
	North Lowther Hills:				
619	Lousie Wood Law	NS 932152	78	E329	
677	Dun Law	NS 917136	78	E329	
732	Green Lowther	NS 900120	78	E329	
725	Lowther Hill	NS 890107	78	E329	
631	East Mount Lowther (Auchenlone)	NS 878100	78	E329	
628	Cold Moss	NS 898094	78	E329	
	South Lowther Hills:				
645	Comb Law	NS 944075	78	E329	
688	Rodger Law	NS 945058	78	E329	
689	Ballencleuch Law	NS 935050	78	E329	
663	Scaw'd Law	NS 922034	78	E329	
672	Wedder Law	NS 938025	78	E329	
668	Gana Hill	NS 954010	78	E329	
672	Wedder Law	NS 938025	78	E329	
611	Earncraig Hill	NS 973013	78	E329	

697   Queensberry   NX 989997   78   E329

40

Culter and Tinto Hills

The old saying has it that 'On Tintock cap there is a mist', which refers to the high mountain of Tinto in Lanarkshire. It rises alone in a wide bend of the River Clyde, and its summit can be seen from Glasgow. Alexander Naismith, famous amongst hill-walkers for his 'Rule', or formula to calculate how long it will take to climb a hill, walked from his home in the city to the summit and back. Close up, the red felsite rock is quite distinctive. The prehistoric cairn on the summit is thought to be one of the largest in Scotland, and it affords extensive views of Clydesdale and to the Culter Hills, which lie to the south-east. There are three main paths to the top—starting to the north, east and south.

The Culter Hills are arranged around the Culter glen, the highest summit being Culter Fell itself. It lies on the east side of the glen, with the two summits of Cardon Hill and Chapelgill Hill to its north-east. Culter Fell can be climbed steeply from the road alongside the Culter Water, either by the shooters' track over Knock Hill or the shepherd's track alongside the King's Beck. Ascents from Glen Holm to the east are less common. I f Culter Fell was 46 feet taller it would qualify as a Corbett.

The summit of Culter Fell is on the boundary between Lanarkshire and Peeblesshire, as are the three summits located further south—Gathersnow Hill, Hillshaw Head and Coomb Dod. The last two now have high tracks over them, giving access to the all-empowering wind turbines that are arranged around the Camps Glen to the west. Off the county boundary, within Peeblesshire, is Coomb Hill.

On the west side of the Culter glen (sometimes spelled Coulter, as in the reservoir, but pronounced 'Cooter' in both cases) rises Hudderstone. In some accounts, this hill is referred to as Heatherstane Law, but current Ordnance Survey maps prefers the former. Its southern slopes are also blighted with turbines and tracks.

41

The Mountains of Great Britain

Height	Name	NGR	OS L	OS E	Ascent
	Tinto Hills:				
711	Tinto	NS 952343	72	E335	
	Culter Hills:				
626	Hudderstone (Heatherstane Law)	NT 022271	72	E336	
635	Coomb Dod	NT 046238	72	E330/E336	
652	Hillshaw Head	NT 048246	72	E336	
688	Gathersnow Hill	NT 059257	72	E336	
640	Coomb Hill	NT 069264	72	E336	
748	Culter Fell	NT 052291	72	E336	
675	Cardon Hill	NT 065315	72	E336	
696	Chapelgill Hill	NT 068303	72	E336	

41

Manor Hills

The Manor Hills rise to the south of Peebles, and spread down to the Megget/Talla gap, which separates them from the Moffat Hills. Most of the hills are rounded, with no great outcrops of rock, but their steep sides and extensive layout make them suitable for strenuous walks. A number of minor roads lead into the hills, that into the Manor valley being the most useful. To its east rise Stob Law, Middle Hill and Dun Rig, collectively the Glenrath Heights. The hills of Deer Law, Black Law and Greenside Heights are readily approached from the Megget dam vicinity to the south.

There are a number of old drove routes across these hills, and the Cross Borders Drove Road long distance path makes its way across the north-eastern hills. Another former drove road is known as the Thief's Road - it traverses The Scrape, Pykestone Hill, Dollar Law and down to the Megget Reservoir. This large sheet of water is totally man-made, the earthen dam being constructed across the glen in 1976.

Cramalt Craig and Broad Law are two sizeable mountains on the north side of the reservoir, and they are usually climbed from the public road hereabouts. Broad Law is the second highest mountain in the Southern Uplands, and qualifies as a Corbett, the highest of only three such summits in the Borders. The easiest means of climbing it is to use the track that leads to the beacon on the summit from Hearthstane, but the ridge from the Megget Stane to the south makes for a more open and unspoilt route.

Most of the other summits in this section are lower, rounded summits, of no major distinction. Deer Law, Conscleuch Head and Greenside Law are only climbed by the keen Donaldist. Talla Cleuch Head, sometimes referred to as Muckle Side, is a steeper summit, usually visited on an ascent of Broad Law. It rises very steeply from the side of the Talla Reservoir, a second man-made loch in the district, constructed in 1905 to supply water to Edinburgh.

The Mountains of Great Britain

Height	Name	NGR	OS L	OS E	Ascent
719	The Scrape	NT 176324	72	E336	
737	Pykestone Hill	NT 173313	72	E336	
668	Drumelzier Law	NT 149312	72	E336	
716	Middle Hill	NT 160294	72	E336	
817	Dollar Law	NT 178278	72	E336	
831	Cramalt Craig	NT 168247	72	E336	
840	△Broad Law	NT 146235	72	E330/E336	
690 est	Talla Cleuch Head	NT 133218	72	E330	
643	Greenside Law	NT 198256	72	E336	
629	Deer Law	NT 223255	73	E337	
698	Black Law	NT 223279	73	E337	
744	Dun Rig	NT 253315	73	E337	
661	Birkscairn Hill	NT 274331	73	E337	
732	Middle Hill	NT 241322	73	E337	
676	Stob Law	NT 230332	73	E337	

42

Moffat & Ettrick Hills

The town of Moffat is a popular location for tourists, though its historical association as a spa town has long gone. One of the spas it was noted for was Hartfell Spa, located on the western slopes of Hart Fell, one of the highest hills in the group, which rises north of the town. It is in excess of 800m but its western slopes are rather rounded and grassy. The more exciting side is that to the east, where the deep glen of Black Hope has a series of rocky craigs along the brae face.

The circle of Black Hope is one of the Borders' finest hillwalks. From Capplegill in Moffat Dale a steep ascent can be made of Nether Coomb Craig, which almost merits inclusion in this list, but whose prominence just misses out by one metre. Swatte Fell is crossed on the way to Hart Fell. This is one of two Corbetts in the Moffat Hills. To the west is the solitary mountain of Whitehope Heights.

The east side of Black Hope is dominated by the twin peaks of Saddle Yoke, Under Saddle Yoke being the higher. To the east side of Saddle Yoke is the Carrifran Glen, the upper slopes ringed by rocky cliffs and Carrifran Gans itself. The glen bottom is occupied by the Carrifran Wood, a rewilding of the area by the Borders Forest Trust, where natural woodland was planted to bring back birch, ash, rowan, juniper and other native trees.

To the north is White Coomb, the highest summit in the section and the second Corbett. Most ascents of this are from the east, using the footpath that passes the Grey Mare's Tail waterfall towards Loch Skeen. The summit is marked by a prehistoric cairn.

North from White Coomb are a series of summits that spread towards the Talla-Megget pass. Of these, Lochcraig Head rises steeply above Loch Skeen, the summit just over 800m. Molls Cleuch Dod is a large rounded summit, only enlivened at the Talla Craigs to the east. It is often ascended from the minor road at Talla Moss to the north.

West of Games Hope is a range of rounded summits stretching north-south from Loch Talla towards Saddle Yoke. Laird's Cleuch Rig is the higher summit south of Garelet Hill, and south again is Erie Hill. There is little to mark the summit of Garelet Dod, though Din Law has a few rock outcrops. Cape Law is another bland summit, basically only discernible by a bend in the fence line.

To the south-east of Moffat Dale rise the Ettrick Hills, so-called because they form an elongated arc around the headwaters of the Ettrick Water. A line of summits lie between the Moffat and Ettrick waters, forming the boundary between Dumfriesshire and Selkirkshire. Herman Law is the northernmost of these, and it only just makes the list on two accounts, being just over 2,000 feet and having a re-ascent of just over 100 feet.

The ridge heads south-west over Andrewhinney Hill, Bell Craig and Mid Hill to Bodesbeck Law. This name was used by James Hogg in his novel, *The Brownie of Bodsbeck*, the Brownie's Cave being located by the Bodesbeck Burn.

The hills to the south are arranged randomly, with Croft Head a steep climb from the Southern Upland Way which passes through the headwaters here, as is Capel Fell. To the south of the long distance path is Loch Fell, the East Knowe being the higher. West Knowe is close to qualifying as another mountain.

Much of the countryside to the east has been planted with extensive conifer forests, one of the largest stretches of such in Britain. Rising above the tree line are a few more mountains—Wind Fell, Hopetoun Craig and Ettrick Pen. These are located on the county boundary, and Ettrick Pen is the highest of the Ettrick Hills group, though its location means that it is not prominent from anywhere significant. Its summit is marked by the remains of a prehistoric cairn. Access to the summit is easiest from Over Phawhope, from where a track leads up part of the western slopes. The Border hills descend in altitude from Ettrick Pen onwards.

Height		Name	NGR	OS L	OS L	OS E	Ascent
		Moffat Hills:					
801	☐	Lochcraig Head	NT 167176	79		E330	
785	☐	Moll's Cleuch Dod	NT 151180	79		E330	
821	☐	△White Comb	NT 163151	79		E330	
757	☐	Carrifran Gans	NT 159138	79		E330	
745	☐	Under Saddle Yoke	NT 142126	78		E330	
722	☐	Cape Law	NT 132150	78		E330	
667	☐	Din Law	NT 124157	78		E330	
698	☐	Garelet Dod	NT 126172	78		E330	
690	☐	Erie Hill	NT 124187	78		E330	
684	☐	Laird's Cleuch Rig	NT 125196	78		E330	
729	☐	Swatte Fell	NT 120115	78		E330	
808	☐	△Hart Fell	NT 113135	78		E330	
637	☐	Whitehope Heights	NT 096139	78		E330	
		Ettrick Hills:					
614	☐	Herman Law	NT 213157	79		E330	

Andrewhinney Hill	NT 197138	79	E330
Bell Craig	NT 187129	79	E330
Mid Rig	NT 180122	79	E330
Bodesbeck Law	NT 170103	79	E330
Smidhope Hill	NT 168076	79	E330
Capel Fell	NT 164069	79	E330
Croft Head	NT 153056	79	E330
Loch Fell	NT 170047	79	E330
Wind Fell	NT 179061	79	E330
Hopetoun Craig	NT 187068	79	E330
Ettrick Pen	NT 200077	79	E330

677	623	616	665	644	678	637	688	665	632	692

Moorfoot Hills

The Moorfoot Hills form a high stretch of upland countryside between Peeblesshire and Midlothian, south of Edinburgh. The rolling countryside is interspersed with deep-sided glens, to the south partially afforested by the Glentress and Leithenwater forests. Glentress Forest is a popular venue for mountain bikers.

Windlestraw Law rises to the east of the B709, Innerleithen to Borthwick Road, a large rounded hill free from forestry. Most ascents will probably be made from Glentress Rig, where a track ascends much of the way. The south-western top, listed by Percy Donald as a Top, almost merits inclusion within this list.

On the west side of the same road rises Whitehope Law, a steep-sided climb from any side. The three remaining Moorfoot summits are best climbed from the north, where they form a circular route around the headwaters of the River South Esk. From near Gladhouse Reservoir a track past Moorfoot follows the South Esk into the glen. A branch ascends The Kipps, from where Blackhope Scar can be climbed. Although on the border, Blackhope Scar is the highest point in Midlothian. The summit is fairly flat and uninteresting, the Garvald Punks and Rough Moss forming an extensive area of boggy peat moss.

The county boundary is then followed westwards to Bowbeat Hill, the upper reaches of which is surrounded by a wind farm, one of the turbines being close to the summit. To the south the landscape is obliterated by Leithenwater Forest, though some tracks can make access easier. The windfarm tracks make walking easier for over a mile, before venturing onto the hillside and heading north to Dundreich, A prehistoric cairn marks the summit, and Jeffries Corse is another Donald Top lying to the north-east, again crowned by a prehistoric cairn. From it, a descent is easily made back towards Gladhouse Reservoir. Dundreich is also often ascended from near Eddleston to the west.

The Mountains of Great Britain

Height	Name	NGR	OS L	OS E	Ascent
☐ 659	Windlestraw Law	NT 371431	73	E337	
☐ 623	Whitehope Law	NT 330446	73	E337	
☐ 651	Blackhope Scar	NT 315483	73	E337	
☐ 626	Bowbeat Hill	NT 292469	73	E337	
☐ 623	Dundreich	NT 274490	73	E337	

245

Cheviot Hills

The Cheviot Hills form the boundary between Scotland and England, the actual border traversing some of the higher summits, which separate the basins of the Tweed and Tyne. There is only one mountain on the border itself—that of Windy Gyle, which is most commonly climbed from Cocklawfoot on the Scottish side, or else from Barrow Burn on the English side. In both cases, ancient drove roads are usually followed to the top, which is surmounted by the prehistoric Russell's Cairn. The Pennine Way also makes its way across the summit on an East-West route, following the border, and detouring to include The Cheviot.

The Cheviot itself is the highest summit in the group, at 815 metres eligible to be a Corbett, if it was on the Scottish side of the border. As such, the summit, an open stretch of boggy ground, is located about a mile into England from the border. The highest point on the border is located below Cairn Hill, at 743 metres, but with insufficient re-ascent to merit listing in its own right.

The other Cheviot summits are fairly undistinguished tops, spread across the moors and surrounded by steep glens which in some cases are afforested. Hedgehope Hill is perhaps the most distinguished of the group, its rounded summit rising above the Hart Hope, and making a prominent hill seen from near Wooler.

Cushat Law and Bloodybush Edge are two rounded hills that only just merit inclusion in this list, the latter just making the 2,000 feet mark, and Cushat Law by little more. Bloodybush Edge has the little distinction of an Ordnance Survey trig point as if to confirm its lowly height. Both of these hills are usually climbed from Bleakhope.

The last summit listed in this section is Cauldcleuch Head, a lonely summit in the middle of the Roxburghshire hills. It too only merits inclusion as a mountain by a few feet.

Height	Name	NGR	OS L	OS E	Ascent
815	⊙The Cheviot	NT 909205	74/75	E16	
714	Hedgehope Hill	NT 944198	80	E16	
652	Comb Fell	NT 924187	80	E16	
615	Cushat Law	NT 928137	80	E16	
610	Bloodybush Edge	NT 902143	80	E16	
619	Windy Gyle	NT 855152	80	E16	
	Roxburgh Hills:				
619	Cauldcleuch Head	NT 457007	79	E331	

North Pennine Hills

The North Pennines Area of Outstanding Natural Beauty occupies the high Pennine Hills around Alston and south towards Brough and Bowes. Much of this countryside comprises of high moorland, though in some areas, particularly the western scarp, these moors plummet considerable to form prominent hills.

To the south-east of Brampton the Tindale Fells climb quite steeply, but only at Cold Fell does the land rise over 2,000 feet, and there only just. The summit is occupied by a prehistoric cairn and a trig point, from where extensive views over the Inglewood Forest and the Vale of Eden to the Cumbrian Mountain can be had. West of Alston are the twin tops of Grey Nag and Black Fell. A third mountain in the vicinity rises to the north of the little village of Renwick, from where a steep path ascends the slopes to the top of Thack Moor, part of the greater Renwick Fell. This summit just reaches 2,000 feet and no more.

South of the Hartside Pass the hills increase in height considerably, with Cross Fell being the highest summit in the Pennine Hills, at 2,930 feet. It is also the highest mountain in England outwith the Lake District. On clear days the Cumbrian Mountains, Southern Uplands and other ranges can be seen. North of Cross Fell are two tops—Melmerby Fell and Fiend's Fell, the former crossed by the ancient Maiden Way Roman road. South of Cross Fell are two shapely hills, Little and Great Dun Fell, and beyond them Knock Fell. All three are traversed by the Pennine Way. Meldon Hill is the last summit in this small area, but its summit is rather bland and flat, and rarely visited.

East of Cross Fell the moors extend over the headwaters of the Tees towards Harwood Common. Round Hill is descriptive of the summit, a place which saw much mining in the past. Like other areas in this section, the slopes are covered with disused shafts and levels. East of Tyne Head, where the great river has its source, is a raised moorland area that rises over 2,000 feet in its centre, where the cairn on Bellbeaver Rigg qualifies as a mountain. To the south-east again rises Herdship Fell, the highest point of which goes by the name of Viewing Hill.

To the south of the impressive High Cup Gill and Maize Beck pass are some more summits, though a number fall within the Warcop danger area. Local directions should be sought prior to heading there. Within the firing range are the Little Fell of Burton Fell and the more attractive Mickle Fell, the highest mountain in County Durham and also in the historical Yorkshire.

On the north-eastern side of Teesdale is a range of hills separating that valley from Weardale. A series of summits here qualify as mountains, although in general they are fairly rounded and flat-topped. The eastmost is a summit that is un-named on Ordnance Survey maps, but which goes by the name of James's Hill. It is easily reached by following the boundary line from the public road at Swinhope Head. Similarly, a walk around Swinhope Moor over Fendrith Hill brings one to Chapelfell Top, like James's Hill, flat and boggy on the summit.

Near the head of Harwood Dale is another flat upland summit, known as High Field, or sometimes Coldberry. The hill is riddled with former mine workings and shake holes. East of Tynehead is Burnhope Seat, and to its north Dead Stones.

To the north of Weardale the high moors continue, with a few points rising above 2,000 feet. Burtree Fell rises due north of Cowshill, again an insignificant top. West of here is Killhope Law, a more rounded summit, more easy to identify the highest part at the trig point. North-east of Nenthead rises The Dodd, which attains just over 2,000 feet. To the south-west of Nenthead a minor road known as the Dowgang Hush makes its way to Garrigill. At the highest point on the road, near to Nunnery Hill, one can walk across two fields to the summit of Flinty Fell, perhaps the easiest mountain to climb in England, if not the whole of Britain.

The Mountains of Great Britain

Height	Name		NGR	OS L	OS E	Ascent
621	Cold Fell		NY 605556	86	E43	
656	Grey Nag		NY 665476	86	E31	
664	Black Fell		NY 648443	86	E31	
610	Thack Moor		NY 611462	86	E31	
614	The Dodd		NY 792458	86/87	E31	
673	Killhope Law		NY 819448	86/87	E31	
612	Burtree Fell		NY 863433	87	E31	
614	Flinty Fell		NY 772419	86/87	E31	
710	Dead Stones		NY 793399	91	E31	
747	Burnhope Seat		NY 785376	91	E31	
708	High Field		NY 824359	91	E31	
651	Three Pikes		NY 834343	91	E31	
703	Chapelfell Top		NY 877347	91	E31	
675	James's Hill		NY 923325	91	E31	
634	Fiend's Fell		NY 643407	86	E31	
709	Melmerby Fell		NY 652380	91	E31	

893	⊙Cross Fell	NY 687343	91	E31
842	Little Dun Fell	NY 704330	91	E31
848	Great Dun Fell	NY 711322	91	E31
794	Knock Fell	NY 721302	91	E31
767	Meldon Hill	NY 781291	91	E19/E31
686	Round Hill	NY 745362	91	E31
620	Bellbeaver Rigg	NY 763351	91	E31
649	Herdship Fell - Viewing Hill	NY 876242	91	E31
675	Murton Fell	NY 754246	91	E19
748	Little Fell	NY 782223	91	E19
788	⊙Mickle Fell	NY 805245	91	E19
619	Blink Moss	NY 875243	91	E19

Northern Lakeland

This group of mountains is located to the north of Keswick, the popular holiday town on Derwent Water. The highest point rises directly north of the town, reaching 931 metres on the peak of Skiddaw, the summit of which is reached by a wide pathway. This begins at a car park at Latrigg, passes a Celtic memorial cross to three local shepherds, surnamed Hawell, from Lonscale farm, and ascends over Jenkin Hill to the summit. Those wishing to bag hill-tops will make the short deviation to Little Man, and perhaps Lonscale Fell. The topmost part of Skiddaw is referred to as Skiddaw Man, at the northern end of the stony ridge. The view from the summit is considerable, including the distant mountains of Galloway (Section 38) to the north, as well as the wider Lake District to the south (if the walker heads to the southern end of the ridge). Another public path to the summit leaves the village of Millbeck and climbs up the Doups to Carl Side and over the western screes. Long Side is another summit that qualifies as a mountain, the ridge west to the Hanging Stone being quite airy. Carl Side is flatter, and it only just merits inclusion, with its re-ascent being 100 feet. An ascent of Skiddaw over Broad End to the north is possible, but rarely done. Skiddaw is composed of Skiddaw Slate, a bluish coloured stone popular in the Lake District.

On Blencathra there is one of Lakeland's better ridges, Sharp Edge, although this is rather short. From Scales a path winds round Mousthwaite Comb to the Glenderamackin valley and to Scales Tarn. Sharp Edge rises to the north of this lake, the edge leading onto the buttresses of Atkinson Pike. The summit of Blencathra is at Hallsfell Top. An easier route to get here is to follow the public path from the Blencathra Centre (formerly a hospital) at the west end and over Blease Fell and along the heads of the gills and tongues. It is the south side of Blencathra that looks the most interesting, the northern slopes being unbroken by rocks and having fairly smooth terrain..

Bowscale Fell and Bannerdale Crags lie to the north-west of Blencathra. Both summits are fairly flat, marked by cairns, but in some places the sides of the hills are more interesting, such as the actual rocks on Bannerdale Crags, or the north side of Bowscale Fell, where there is a deep drop over the Tarn Crags to Bowscale Tarn, a classic glacial lake. A circuit round Bannerdale can be made to climb these hills. This starts and ends at Mungrisdale village. Within the steep cliffs of Bannerdale Crags are various lead workings and a cave. The circuit is better if Souther Fell is included, even although it does not top 2,000 feet.

Four other summits in this range of hills are over the magical height. Carrock Fell rises to the north-west of Mosedale, its lower slopes the haunt of rock-climbers. The summit is topped by an Iron-age fort. Alfred Wainwright reckoned that this was the second most interesting summit in the northern Cumbrian hills, after Blencathra. This may be due to the fact that the hill comprises of gabbro, a rough igneous rock. High Pike can be reached from Caldbeck by means of a track past an old mine to the Curly Job Well, forming part of the Cumbria Way. From the well the summit is a short climb to the south. The way can be continued over Great Lingy Hill to Knott, another mountain, and a return down to the old mines at the head of the Dale Beck, from where a path returns to Fell Side and on to Caldbeck. Great Calva is more remote, but is best climbed from the west, a track from near Bassenthwaite village to Skiddaw House (another part of the Cumbria Way) passing the western slopes of the hill.

Apart from the Skiddaw group of Long Side, Carl Side and Lonscale Fell, and Blencathra, the rest of the hills are sufficiently remote to allow the walker to spend his time away from the crowds which can be a problem on so many of Lakeland's popular peaks. These remoter summits were the haunt of John Peel, the famous huntsman of the song, whose grave is located in Caldbeck churchyard.

47

The Mountains of Great Britain

Height	Name	NGR	OS L	OS E	Ascent
	Caldbeck Fells:				
661	Carrock Fell	NY 341337	90	E5	
658	High Pike	NY 319350	90	E5	
710	Knott	NY 296330	89/90	E4	
	Skiddaw Forest:				
690	Great Calva	NY 291312	89/90	E4	
931	● Skiddaw	NY 260291	89/90	E4	
746	Carl Side	NY 254282	89/90	E4	
734	Longside Edge	NY 249285	89/90	E4	
865	Little Man (Low Man)	NY 267278	89/90	E4	
715	Lonscale Fell	NY 285271	89/90	E4	
868	⊙ Saddleback (Blencathra)	NY 323177	90	E5	
683	Bannerdale Crags	NY 335290	90	E5	
702	Bowscale Fell	NY 333306	90	E5	

Western Lakeland

The area covered by this section extends to the west of the Bassenthwaite, Keswick, Thirlmere, Ambleside gap in the Cumbrian Mountains. Many of the ranges form something akin to spokes, radiating from a hub around Rosthwaite in Borrowdale. The most northerly group of mountains are to the south of the Whinlatter Pass, west of Keswick, where Grisedale Pike forms a prominent cone when seen from near Keswick. West of it rises the lower but more rugged Hopegill Head and Whiteside, rising steeply above Lorton Vale.

To the south of Gasgale is the highest summit in this northerly group—Grasmoor, which reaches 2,795 feet, high above Crummock Water. Adjoining Grasmoor is Crag Hill, and from it is a ridge of lesser tops descending to Causey Pike, another distinctive top seen from Derwent side.

To the south of Newlands Hause are the Derwent Fells proper, a series of attractive mountains including Robinson, Hindscarth and Dale Head, with High Spy further east.

To the south-west of Buttermere are the mountains that form an elongated ring around Ennerdale. The north side starts at Great Borne to the west, the rounded hill just rising over 2,000 feet. From there the range includes Starling Dodd, Red Pike and the highest point, High Stile, an English Corbett. On the south side of the valley are Iron Crag, Haycock, Little Scoat Fell and Black Crag, before the summit of Pillar is reached.

Situated at the heads of Ennerdale, Wasdale and Borrowdale are the mountains of Great Gable, Green Gable and Kirk Fell, with a few lesser summits around them. At 2,949 feet, Great Gable is one of the higher mountains in the area.

The Borrowdale Fells are located around the headwaters of the Derwent and Langstrath watercourses. Ullscarf and High Raise are two prominent hills on the east side of Langstrath, but to the south of here are the Langdale Pikes, the Pike of Stickle and Harrison Stickle being prominent rock outcrops.

In the midst of these mountain groups is the range that contains Scafell Pike, England's highest mountain at 3,209 feet, plus the nearby Sca Fell, Broad Crag and Ill Crag, all of which attain the 3,000 feet line. Prior to accurate surveying, Sca Fell was thought to have been the tallest mountain in England. Great End, although over 900 metres, fails to reach 3,000 feet. Another summit that just fails to reach a magic contour line is Illgill Head, south-east of Wast Water, which is only 609 metres tall, and thus fails to qualify for this list.

North of the Wrynose Pass is a series of hills that links the pass with Scafell Pike. Here are a series of prominent summits, rising over Long Top and Crinkle Crags to Bow Fell, one of the district's taller mountains, at 2,959 feet. It is sometimes spelled Bowfell. Alfred Wainwright reckoned it was one of the best six mountains in the Lake District.

The final group of mountains in this section are the Furness Fells, which lie south of Wrynose Pass and west of Coniston village. Of these the Old Man of Coniston is the tallest, a prominent mountain rising steeply above the village of the same name. Much of its slopes were formerly quarried. It was the highest point of Lancashire before county lines were redrawn, though Swirl How gives it a good run for its money. West of it is the attractive Dow Crag. Further north are the peaks of Grey Friar, Swirl How and Wetherlam, which afford great walks. Harter Fell is an outlier, to the south of Eskdale.

Many of the mountains in this section are owned by the National Trust, and all of them fall within the Lake District National Park. Scafell Pike and surroundings were presented to the Trust by Lord Leconfield as a war memorial to Lake District soldiers. Similarly, Great Gable and 3,000 acres were purchased for a similar reason by the Fell & Rock Climbing Club, the lands also donated to the National Trust.

The Mountains of Great Britain

Height	Name	NGR	OS L	OS E
☐ 719	Whitesdide	NY 175221	89/90	E4
☐ 770	Hopegill Head (Hobcarton Pike)	NY 185221	89/90	E4
☐ 739	Longcrag Pike	NY 193220	89/90	E4
☐ 791	⊙ Grisedale Pike	NY 198226	89/90	E4
☐ 852	⊙ Grasmoor	NY 175202	89/90	E4
☐ 839	Crag Hill (Eel Crag)	NY 192203	89/90	E4
☐ 773	Sail	NY 198203	89/90	E4
☐ 672	Scar Crags	NY 209207	89/90	E4
☐ 637	Causey Pike	NY 217209	89/90	E4
☐ 660	Whiteless Pike	NY 180189	89/90	E4
	Derwent Fells:			
☐ 737	Robinson	NY 202168	89/90	E4
☐ 727	Hindscarth	NY 216165	89/90	E4
☐ 753	Dale Head	NY 223153	89/90	E4
☐ 653	High Spy (Scawdel Fell)	NY 233162	89/90	E4
☐ 616	Great Borne	NY 123163	89	E4/E303

Ascent

Height	Name	NGR	OS L	OS E	Ascent
633	Starling Dodd	NY 142157	89	E4	
755	Red Pike	NY 160154	89	E4	
807	⊙ High Stile	NY 170148	89	E4	
744	High Crag	NY 180140	89/90	E4	
648	Fleetwith Pike	NY 206141	89/90	E4	
715	Brandreth	NY 215119	89/90	E4	
646	Base Brown	NY 225114	89/90	E4	
801	Green Gable	NY 215107	89/90	E4/E6	
899	⊙ Great Gable	NY 211102	89/90	E4/E6	
787	East Kirk Fell	NY 199108	89/90	E4/E6	
802	⊙ Kirk Fell	NY 194104	89/90	E4/E6	
892	⊙ Pillar	NY 171121	89/90	E4	
828	Black Crag	NY 166126	89	E4	
841	Scoat Fell - Little Scoat Fell	NY 160113	89	E4	
826	Red Pike	NY 166106	89	E4/E6	
616	North Yewbarrow (Stirrup Crag)	NY 176092	89/90	E4/E6	

48

627	☐	Yewbarrow	NY 173084	89/90 E6
797	☐	Haycock	NY 144107	89 E4/E6
640	☐	Iron Crag	NY 121122	89 E4/E303
692	☐	Seatallan	NY 140084	89 E5/E303
631 est	☐	Rosthwaite Fell	NY 256113	89/90 E4
781	☐	Glaramara	NY 247105	89/90 E4/E6
721	☐	Lincomb Hill	NY 243097	89/90 E4/E6
785	☐	Allen Crags	NY 237085	89/90 E6
632	☐	Seathwaite Fell	NY 227097	89/90 E4/E6
910	☐	Great End	NY 227083	89/90 E6
935	☐●	Ill Crag	NY 223083	89/90 E6
934	☐●	Broad Crag	NY 227083	89/90 E6
807 est	☐	Lingmell	NY 209082	89/90 E6
978	☐●	Scafell Pike	NY 216027	89/90 E6
964	☐●	Sca Fell	NY 207065	89/90 E6
726	☐	Ullscarf	NY 292122	89/90 E4
762	☐⊙	High Raise	NY 281095	89/90 E4/E6

48

48

Height	Name	NGR	OS L	OS E	Ascent
	Langdale Fell:				
723	Thunacarr Knott	NY 279080	89/90	E6	
736	Harrison Stickle	NY 282073	89/90	E6	
709	Pike of Stickle	NY 273073	89/90	E6	
651 est	Rossett Pike	NY 249075	89/90	E6	
885	Esk Pike	NY 237075	89/90	E6	
902	Bow Fell	NY 245064	89/90	E6	
859	Crinkle Crags (Long Top)	NY 248049	89/90	E6	
834	Crinkle Crags (South Top)	NY 250045	89/90	E6	
701	Cold Pike	NY 263035	89/90	E6	
705	Pike of Blisco	NY 271042	89/90	E6	
	Furness Fells:				
773 est	Grey Friar	NY 260004	89/90	E6	
802	Swirl How	NY 272004	89/90	E6	
745	Black Sails	NY 283007	89/90	E6	
762	Wetherlam	NY 288011	89/90	E6	

48

⊙ Old Man of Coniston	SD 272978	96/97 E6
Dow Crag	SD 262978	96/97 E6
Harter Fell	SD 218997	96 E6

☐ 803
☐ 778
☐ 653

Eastern Lakeland

The Eastern Cumbrian mountains extend from the A591 road, which passes from Keswick to Windermere, eastwards over the Kirkstone Pass towards Shap. The group is divided into two main ranges, the Helvellyn group and the High Street group.

The northernmost summit in this section is Clough Head, a prominent hill overlooking the Greta valley to the east of Keswick. From it the summits of Matterdale Common spread southwards, gradually increasing in height until Helvellyn is reached, an English Munro. Some of the intervening hills are attractive climbs, such as Stybarrow Dod and Sheffield Pike. This part of the hills forms Matterdale Common, which is owned by the National Trust.

Helvellyn is England's third highest peak, making it a popular climb. The most attractive routes are from the east, where the ridges of Catstye Cam and Striding Edge provide airy access routes. Dollywaggon Pike merits inclusion in the list as a mountain in its own right, the southernmost summit of the Helvellyn ridge. Ascents from the west are more straightforward, being simply steady ascents from Thirlmere side, with popular starting points at Wythburn to the south end of the lake, or from Highpark Wood at Helvellyn Gill.

Helvellyn has an interesting history. On the summit flat is a memorial commemorating the landing on the mountain (and subsequent take off) of an Avro 585 aeroplane in 1926. The artist Charles Gough was killed on Striding Edge in 1805 and a second memorial commemorates him. A cross on Striding edge marks the spot where Robert Dixon fell whilst running with the Ullswater Hounds in 1858.

Beyond Grisedale the bulk of Fairfield rises, an attractive mountain with steep rocky coves on various sides. Great Rigg and Hart Crag are substantial shoulders of Fairfield, meriting inclusion as separate mountains. Most ascents are made from near Grasmere. North-east of Fairfield is The Cape, the summit fairly flat but having steep sides. South-east of Fairfield are the summits of Dove Crag, and Red Screes, the latter summit quickly ascended from the highest point of the Kirkstone Pass.

The ancient Roman Road of High Street makes its way south over the mountains from Ullswater, linking the ancient forts at Brougham and Ambleside. Its first mountain is Loadpot Hill, followed by High Raise and Rampsgill Head, before reaching the mountain that bears the road's name—High Street. The flat summit also bears the name Racecourse Hill, for here fairs were held on 12 July in the eighteenth and nineteenth centuries, the last recorded one being held in 1835. Stray sheep were swapped, and various sporting events took place. Perhaps one of the finest routes to the summit is that over Rough Crag and Long Stile, a simpler version of Helvellyn's Striding Edge. Away to the north of High Street is Place Fell, the dominant hill at the east side of Patterdale at the head of Ullswater.

49

To the south of High Street is a complicated selection of mountains, with valleys and ridges interspersed with random tarns. Stony Cove Pike rises to the west, a flat-topped mountain rising east of the Kirkstone Pass. This mountain, and the west side of High Street, form another block of countryside protected by the National Trust. The old High Street skirts the summit of Thornthwaite Crag, descending into the Hagg Gill valley, avoiding the more pointed summits of Froswick, Ill Bell and Yoke. The summit of Ill Bell is dominated by a number of large cairns.

Harter Fell rises at the headwaters of the River Kent and above Haweswater. The ridge to the south includes Kentmere Pike, a high summit ascended fairly easily from Kentmere itself.

The hills to the east of here are much quieter, being lower in height and less rocky and rugged in appearance. The list includes Branstree, Tarn Crag and Sleddale Fell. Selside Pike has a prehistoric cairn on its summit. The hills diminish in size and bulk as they drop to form the Shap Fells.

The Mountains of Great Britain

Height	Name	NGR	OS L	OS E	Ascent
726	Clough Head	NY 333225	90	E5	
857	Great Dodd	NY 342204	90	E5	
843	Stybarrow Dod	NY 341187	90	E5	
795	White Stones	NY 353188	90	E5	
675	Sheffield Pike	NY 369181	90	E5	
883	Raise	NY 343173	90	E5	
863	White Side Bank	NY 338167	90	E5	
890	Catstye Cam (Catchedicam)	NY 348158	90	E5	
950	●Helvellyn	NY 341151	90	E5	
858	Dollywaggon Pike	NY 346131	90	E5	
736	Seat Sandal	NY 343115	90	E5	
841	⊙The Cape (St Sunday Crag)	NY 369134	90	E5	
873	⊙Fairfield	NY 359118	90	E5	
766	Great Rigg	NY 356104	90	E5/E7	
822	Hart Crag	NY 369112	90	E5	
792	Dove Crag	NY 374104	90	E5/E7	

☐	637	Little Hart Crag	NY 387100	90	E5/E7
☐	776	⊙ Red Screes	NY 396088	90	E7

Martindale Common:

☐	672	Loadpot Hill	NY 457181	90	E5
☐	802	High Raise	NY 448134	90	E5
☐	792	Rampsgill Head	NY 443128	90	E5
☐	696	Rest Dodd	NY 432137	90	E5
☐	657	Place Fell	NY 405169	90	E5
☐	763	⊙ Stony Cove Pike	NY 418100	90	E5/E7
☐	784	Thornthwaite Crag	NY 431100	90	E5/E7
☐	720	Froswick	NY 435085	90	E7
☐	757	Ill Bell	NY 437077	90	E7
☐	706	Yoke	NY 438068	90	E7
☐	828	⊙ High Street	NY 441111	90	E5
☐	628	Rough Crag	NY 454112	90	E5
☐	778	Harter Fell	NY 461094	90	E5/E7
☐	730	Kentmere Pike	NY 465068	90	E7

49

49

Height		Name	NGR	OS L	OS E	Ascent
655	☐	Selside Pike	NY 491112	90	E5	
713	☐	Artlecrag Pike (Branstree)	NY 478100	90	E5/E7	
664	☐	Tarn Crag	NY 488078	90	E7	
638	☐	Sleddale Fell (Grey Crag)	NY 497072	90	E7	

South Pennine Hills

Most of the mountains in this section are located within the Yorkshire Dales National Park. The hills stretch from the A66 trans-Pennine road southwards to the Harrogate-Skipton-Settle line.

The Howgill Fells occupy the furthest north-west part of this section, a series of steep hills climbing north of Sedbergh and east of the M6. Five of these summits rise above 2,000 feet, The Calf being the tallest. It is traversed by the Dales High Way. Calders to the south of The Calf just qualifies as a mountain, whereas the other tops are more prominent, if lower, tops.

East of Sedbergh are three tops that have steep sides but are rather flat-topped; Wild Boar Fell, Swarth Fell and East Baugh Fell. Great Knoutberry Hill is the highest point of Widdale Fell.

West of Barbondale rises Calf Top, which was for many years thought to be only 1999 feet in height. Recent surveys have confirmed that it just reaches the 2,000 foot contour.

Great Coum and Gragareth form two ends of a long ridge separating Kingsdale from Barbondale. The former gets its name from the Great Combe, or corrie, on its eastern slopes, the top of the hill being slightly taller than the nearby Crag Hill. East of Kingsdale is Whernside, a prominent and distinctively-shaped summit, with a steep eastern side. It is Yorkshire's tallest mountain and one of the 'Three Peaks' ascended in a race.

To the east of Ingleton rises the prominent Ingleborough, noted for its steep sides with horizontal stone banding of limestone. This is Yorkshire's second-tallest mountain and another of the 'Three Peaks' The surroundings have been designated a National Nature Reserve. The summit of Ingleborough is crowned by an extensive Iron Age hill fort. Adjoining the hill is a second mountain, Simon Fell, which has a less distinguished summit, being fairly flat over an extensive area, but with steep sides.

The final 'Three Peaks' summit is Pen-y-ghent, which rises on the east side of Ribblesdale. The summit is fairly conical, with steep sides on most approaches, apart from the north, where the hill is conjoined to Plover Hill. The Pennine Way traverses the summit. A wide ridge from Pen-y-ghent swings around the head of Littondale to Birks Fell, a summit that only just reaches 2,000 feet.

South-east of the minor road that crosses from Stainforth to Halton Gill are two mountain summits— Fountains Fell and Darnbrook Fell. Fountains Fell has two flattish summits, the northern one being the higher. It is covered with old mine workings. The Pennine Way skirts past the highest point, but keen mountain climbers will make a detour to tick off the summit on their lists. Much of the south and east side of these hills is owned by the National Trust.

To the south of the village of Hawes a minor road climbs up Sleddale to join a high route that was originally a Roman road. Off this road to the east is Drumaldrace, a low mountain topped by a cairn. North of the old Roman Road is Dodd Fell Hill, as the highest point of Dodd Fell is known. This name is something of a tautological hybrid, for 'dodd', 'fell' and 'hill' basically all mean the same thing.

In Langstrothdale Chase is a summit that rises over 2,000 feet but which has a flat and boggy summit. The highest point is marked by a trig point, and the hill goes by the name of Middle Tongue, though this probably only officially describes the low ridge between Thornrake Gill and Middle Tongue Gill. Other lists refer to the top as Yockenthwaite Moor.

On the east side of Wharfedale are Great Whernside and Buckden Pike, two similarly tall mountains.

At the head of Swaledale a series of summits rise up in the moors. Great Shunner Fell and High Seat are the best two summits in this small group.

The Mountains of Great Britain

Height	Name	NGR	OS L	OS E	Ascent
	Howgill Fells:				
624	Randygill Top	NY 687000	91	E19	
639	Yarlside	SD 685985	98	E19	
676	The Calf	SD 667971	98	E19	
674	Brant Fell (Calders)	SD 671961	98	E19	
641 est	Fell Head	SD 649982	97	E19	
708	Wild Boar Fell	SD 758988	98	E19	
681	Swarth Fell	SD 755967	98	E19	
678	Tarn Rigg Hill (East Baugh Fell)	SD 741917	98	E19	
672	Great Knoutberry Hill	SD 788871	98	E2	
736	Whernside	SD 738814	98	E2	
687	Great Coum	SD 700836	98	E2	
627	Gragareth (Greygarth Hill)	SD 688793	98	E2	
610	Calf Top	SD 664856	98	E2	
724	Ingleborough Hill	SD 741746	98	E2	
650	Simon Fell	SD 755752	98	E2	

50

The Mountains of Great Britain

Height	Name	NGR	OS L	OS E	Ascent
	Littondale Hills:				
668	Fountains Fell	SD 864715	98	E2/E30	
624	Darnbrook Fell	SD 884728	98	E30	
694	Pen-y-ghent	SD 838734	98	E2	
682 est	Plover Hill	SD 849752	98	E2/E30	
610	Birks Hill	SD 919763	98	E30	
668	Dodd Fell Hill	SD 841845	98	E2/E30	
614	Drumaldrace (Wether Fell)	SD 873866	98	E2/E30	
643	Middle Tongue (Yockenthwaite Moor)	SD 909811	98	E30	
702	Buckden Pike	SD 961788	98	E30	
704	Great Whernside	SE 002739	98	E30	
	Swaledale Hills:				
675	Lovely Seat	SD 879951	98	E19/E30	
716	Great Shunner Fell	SD 848973	98	E19/E30	
666	Sails (Lunds Fell)	SD 808966	98	E19	
709	High Seat	NY 802012	91	E19	

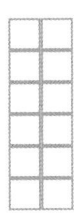

Nine Standards Rigg	NY 825062	91	E19
Rogan's Seat	NY 919031	91	E30

662 Nine Standards Rigg NY 825062 91 E19
672 Rogan's Seat NY 919031 91 E30

50

High Peak

The High Peak refers to the upland area between Sheffield and Stockport, part of the Peak District National Park (the first National Park in Britain). Although elevated and part of the Pennine Hills, there are only two points where the land rises in excess of 2,000 feet, Kinder Scout and Bleaklow Head.

Kinder Scout is the name of an extensive upland moor, a great stretch of flat moorland surrounded by steep sides. A fair area of the moor is over 2,000 feet, but as the land is fairly flat, only one of the low summits qualifies as a mountain. The specific spot, however, is undistinguished and often walkers will ignore it in preference for Kinder Low, a slight eminence with a trig point, cairn and which is traversed by the Pennine Way. The actual highest point lies half a mile to the north-east of Kinder Low, nothing on the ground marking the exact spot. It is Derbyshire's highest point.

Kinder Scout is famed in hillwalking history for the mass trespass that took place there in 1932. The landowners at the time refused to allow ramblers onto their ground, resulting in the protest, which was one of many. Much of Kinder Scout was acquired by the National Trust allowing open access. Kinder Scout is also a national nature reserve.

Bleaklow Head is the highest point of a similar upland plateau to the east of Glossop known as Bleaklow. The highest point is located at the Wain Stones, a point crossed by the Pennine Way on its way through the backbone of England. A cairn a few hundred yards to the north-east is at a similar height, and Bleaklow is the farthest east stretch of countryside in the United Kingdom that rises above 2,000 feet. Like Kinder Scout, Bleaklow is part of the National Trust's High Peak Moors property. In 1948 a USAF Boeing Superfortress named 'Overexposed' crashed on the hill, killing all thirteen on board. A small memorial can be found at Higher Shelf Stones.

51

Height	Name	NGR	OS L	OS E	Ascent						
☐ 636	Kinder Scout	SK 087875	110	E1							
☐ 633	Bleaklow Head	SK 092958	110	E1							

51

North Gwynedd

The boundary of this section is formed on the south by the A498 and A4086. All the mountains of the North Gwynedd section fall within the Snowdonia National Park, the third such park to be established in Britain, covering over 800 square miles. The park takes its name from the highest peak in England and Wales, Snowdon, or Yr Wyddfa in Welsh, which rises 1,085 metres in height and is the most popular mountain to ascend in Britain. Many tourists reach the summit by one of the seven paths which climb from all directions, or else by taking the Snowdon Mountain Railway. This climbs the peak in a five mile ride from Llanberis, and was first opened in 1896. One of the most spectacular ascents of the mountain is the route round the Llyn Llydaw horseshoe. Beginning at Pen-y-pass, a path climbs up the Grib-goch ridge to Crib-y-ddysgl. The way is made alongside the railway to the summit of Snowdon, then a descent over Bwlchysaethau to Y Lliwedd. From here one follows the path down the ridge to Llyn llydaw and the Miner's Track back to Pen-y-pass. An interesting walk can be had from Snowdon north-westwards over Moel Cynghorion to Moel Eilio and a descent to Llanberis.

Fifteen of the mountains in this district qualify as 'Furths', that is the mountains in England and Wales that would be classed as a Munro, if in Scotland. All of Wales' Furths fall within this section. Five of these summits are in excess of 1,000m, making them some of the highest mountains in Britain.

North of the Snowdon group, across the Pass of Llanberis, is the Glyder group, Glyder Fawr just managing to make the 1,000 metre level after some resurveying. Rounded to the south, these mountains are much more spectacular from the Nant Ffrancon side where they display mile upon mile of rock buttresses. In this group is Wales' most spectacular mountain, Tryfan, a narrow projection of rock between two cwms, its summit located on a tiny pinnacle. From

52

Llyn Ogwen a steep climb by the side of the Milestone Buttress brings one to the summit.

A traverse of all the Glyders in a single day makes great walking. From Helyg an ascent is made of Gallt yr Ogof followed by Y Foel Goch and Glyder Fach. The stony summit is crossed en route to Glyder Fawr and a sharp descent to Llyn y Cwn. A path drops steeply down into the Devil's Kitchen if an escape is needed to the north. Otherwise climb up Y Garn and Foel-goch. If time allows an ascent of Elidir Fawr can be included before doubling back to cross Carnedd y Filiast on the descent to Nant Ffrancon. This gives a walk of eleven miles.

All of the northern slopes of the Glyders are owned by the National Trust, as are the Carneddau, on the other side of Llyn Ogwen, Carnedd Llywelyn reaching 1,062 metres. This formed part of Penrhyn Castle estate. From Llyn Ogwen a path climbs Penyrole-wen and continues to Carnedd Dafydd. A further walk can be made to Carnedd Llywelyn and a descent over Penyrhelgi-du to Tal-y-braich. This brings one out onto the A5 three miles east of the starting point. Further north the hills drop in height to Taly y Fan at 610 metres and the rock headland of Penmaen Bach, the public road having to force its way through a tunnel. From Llanfairfechan a track climbs up to the summit of Drum, crowned with the ancient Carnedd Penydorth-goch. A path along the ridge continues over Foel-fras, Garnedd-uchaf and Foel-grach to Carnedd Llywelyn. The return route can strike off at Foel-grach and descend over Drosgl, with its Bronze Age cairn, to the Aber Falls and back to Llanfairfechan. Total distance of the walk is seventeen miles, though the journey out to Llywelyn may be missed out.

South-west of the Snowdown group are eight summits which offer quieter walking, only Moel Hebog attracting climbers in any numbers, a path scaling the hill from Beddgelert. These hills afford extensive views of their higher neighbours, and westwards across the Irish Sea.

Height	Name	NGR	OS L	OS E	Ascent
610	Tal y Fan	SH 729727	115	E17	
770	Drum	SH 708696	115	E17	
849	Llwytmor	SH 687693	115	E17	
942	●Foel-fras	SH 697682	115	E17	
758	Drosgl	SH 663680	115	E17	
926	●Carnedd Gwenllian (Carnedd Uchaf)	SH 687669	115	E17	
976	●Foel Grach	SH 688659	115	E17	
962	●Yr Elen	SH 672651	115	E17	
1064	●Carnedd Llywelyn	SH 684644	115	E17	
1044	●Carnedd Dafydd	SH 663630	115	E17	
978	●Pen yr Ole Wen	SH 656620	115	E17	
833	Pen yr Helgi Du	SH 698630	115	E17	
799	◉Pen Llithrig y Wrach	SH 715623	115	E17	
678	Creigiau Gleision	SH 729615	115	E17	
634	Creigiau Gleision Bach	SH 733622	115	E17	
821	Carnedd y Filiast	SH 620628	115	E17	

52

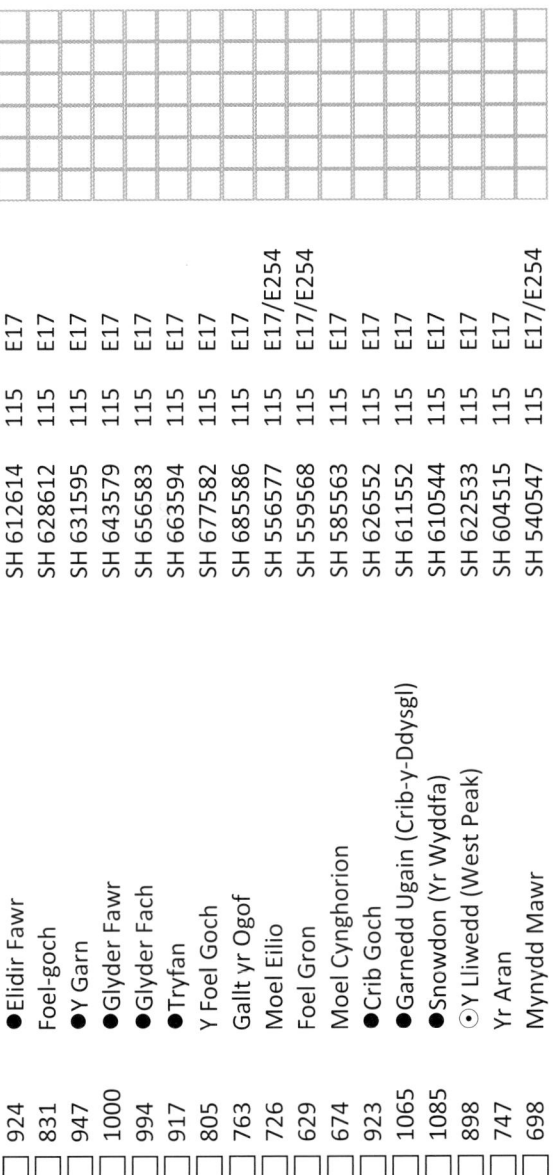

924	● Elidir Fawr	SH 612614	115	E17
831	Foel-goch	SH 628612	115	E17
947	● Y Garn	SH 631595	115	E17
1000	● Glyder Fawr	SH 643579	115	E17
994	● Glyder Fach	SH 656583	115	E17
917	● Tryfan	SH 663594	115	E17
805	Y Foel Goch	SH 677582	115	E17
763	Gallt yr Ogof	SH 685586	115	E17
726	Moel Eilio	SH 556577	115	E17/E254
629	Foel Gron	SH 559568	115	E17/E254
674	Moel Cynghorion	SH 585563	115	E17
923	● Crib Goch	SH 626552	115	E17
1065	● Garnedd Ugain (Crib-y-Ddysgl)	SH 611552	115	E17
1085	● Snowdon (Yr Wyddfa)	SH 610544	115	E17
898	⊙ Y Lliwedd (West Peak)	SH 622533	115	E17
747	Yr Aran	SH 604515	115	E17
698	Mynydd Mawr	SH 540547	115	E17/E254

52

52

Height	Name	NGR	OS L	OS E	Ascent
610	Mynydd Graig Goch	SH 497485	115	E17/E254	
734	Craig Cwm Silyn	SH 526504	115	E17/E254	
653	Mynydd Tal-y-mignedd	SH 535514	115	E17/E254	
709	Trum y Ddysgl	SH 545516	115	E17/E254	
695	Mynydd Drws-y-coed	SH 549519	115	E17/E254	
638	Moel Lefn	SH 553485	115	E17/E254	
655	Moel yr Ogof	SH 556478	115	E17/E254	
782	⊙Moel Hebog	SH 565469	115	E17/E254	

Central Gwynedd

The mountains around Blaenau Ffestiniog form the basis of this section. The northern boundary is the road from Betws-y-Coed westwards through Beddgelert to Porthmadog. To the south the boundary is the Dolgellau to Bala valley.

Virtually all of the summits are located within the Snowdonia National Park and access is generally easy. At the north end rises Carnedd Moel Siabod, a prominent summit that merits English Corbett status. It is often climbed from Capel Curig, from where it forms a prominent summit on the south-western horizon, there being a variety of paths heading to the ridge of Moel Siabod.

The next summit to rise above 2,000 feet is Ysgafell Wen, located amid a rough stony countryside spattered with little tarns. The highest point on this summit is the southern boss, the one to the north with the cairn being lower. To the north is Moel Meirch, a hill that just fails to be included in this list, being only 609 m in height.

Moel Terfyn and Moel Druman follow, two summits similar to Ysgafell Wen in that they form complicated rounded blocks of rock which are difficult to identify as being the highest point. Allt-fawr is a different style altogether, its summit being considerably higher and its steep eastern and southern sides making it a prominent mountain above the Ffestiniog Slate Quarry at Blaenau Ffestiniog.

The summits of Manod Mawr rise due east of Blaenau Ffestiniog, separated by the Graig-ddu Quarry. This prevents a direct walk from one summit to the other, the works usually avoided by walking around them to the west. The north-east summit is being encroached upon by the slate quarry, and unless it is preserved may become a summit that is lost in the future.

Cnicht, Moel-yr-hydd, Moelwyn Mawr and Moelwyn Bach have more obvious summits than the moels west of Allt-fawr and make popular walks from the various valleys below. The two Moelwyns are probably the most popular, being prominent west of Blaenau Ffestiniog and there being a series of public rights of way over the shoulders, originally used by the slate quarriers. The east side of Moelwyn Mawr drops steeply into the Llyn Stwlan reservoir, used to generate electricity at the power station below. Cnicht is usually climbed from Croesor to the south west, a popular path making its way up the long ridge to the summit.

Between the coastal A496 and the A470 is a range of hills usually referred to as the Rhinogs, from two summits known as Rhinog Fawr and Rhinog Fach. These two rocky mountains are almost comparable in height, but Rhinog Fawr looks larger in bulk, resulting in it gaining the 'big' appellation in Welsh. The south side of Rhinog Fawr and the east side of Y Llethr is National Trust property, as are one or two other random stretches of countryside. Rhinog Fawr is a National Nature Reserve. The most northerly summit in this part of the section is Moel Ysgyfarnogod, often reached from Eisingruig to the west, or else from near Trawsfynydd to the east. Foel Penolau just fails to merit inclusion as a mountain due to the re-ascent rule.

South of the Rhinog summits rises Diffwys, a major mountain with a rocky east face. This is often ascended from the south, paths to the summit leaving from near Bontddu. Y Garn to the east is a less-exciting and thus less-popular summit, quickly climbed from Blaen-y-cwm.

East of the A470 and west of Bala are the Arenig hills. Carnedd y Filiast lies to the north, a rounded summit with a path to it from the south. The two Arenigs are more interesting. Arenig Fach is north of the A4212, a solitary dome crowned by the ancient Carnedd y Bachgen and with a rocky outcrop overlooking Llyn Arenig Fach. Arenig Fawr is much taller, is also crowned by a prehistoric cairn, and with Llyn Arenig Fawr at its feet. A series of lower summits spread south-west, until the more prominent Rhobell Fawr is reached.

53

The Mountains of Great Britain

Height	Name	NGR	OS L	OS E	Ascent
872	⊙ Carnedd Moel-siabod	SH 705547	115	E18	
669	Ysgafell Wen	SH 663485	115	E17	
672	Moel Terfyn	SH 667481	115	E17	
689	Cnicht	SH 645466	115	E17	
648	Moel-yr-hydd	SH 672454	115	E17/E18	
770	⊙ Moelwyn Mawr	SH 658449	124	E18	
710	Moelwyn Bach	SH 660438	124	E18	
698	Allt-fawr	SH 682474	115	E17	
676	Moel Druman	SH 671476	115	E17	
623	Moel Penamnen	SH 717483	115	E18	
658	Manod Mawr (N.E. Top)	SH 728458	115	E18	
661	Manod Mawr (S.W. Top)	SH 724447	124	E18	
623	Moel Ysgyfarnogod	SH 658346	124	E18	
720	Rhinog Fawr	SH 656290	124	E18	
712	Rhinog Fach	SH 665270	124	E18	
756	Y Llethr	SH 661258	124	E18	

53

53

Height	Name	NGR	OS L	OS E	Ascent
750	Diffwys	SH 661234	124	E18	
629	Y Garn	SH 703230	124	E18	
689	Arenig Fach	SH 831415	124/125	E18	
611	Foel Goch	SH 953423	125	E18	
669	Carnedd y Filiast	SH 871446	124/125	E18	
854	⊙ Arenig Fawr	SH 827369	124/125	E18	
751	Moel Llyfnant	SH 808352	124/125	E18/E23	
619	Foel Boeth	SH 778345	124	E18/E23	
662	Dduallt	SH 811273	124/125	E23	
734	Rhobel Fawr	SH 786257	124	E23	

South Gwynedd & North Powys

The Berwyn range extends across the borders of northern Powys, Gwynedd, Denbighshire and even part of Wrexham. Moel Fferna is the mountain nearest Llangollen, its prehistoric-cairn crowned summit just in Denbighshire. Berwyn extends south-west from here, over Moel Gwyn to Cadair Bronwen, again topped by a prehistoric burial cairn known as Bwrdd Arthur.

Cadiair Berwyn is the highest mountain in this group, the highest point actually the southern top, not where the ancient cairn and trig point are located. With Moel Sych, this stretch of Berwyn makes interesting and airy walking. Parts of the slopes form part of Y Berwyn National Nature Reserve. The border between Wrexham and Denbighshire has two summits on it, Foel Wen and Mynydd Tarw. Moel Sych's long south-westerly ridge curls round the headwaters of the Afon Disgynfa, often remaining above 2,000 feet, but only Post Gwyn qualifies as a separate mountain.

West of Pont Cwm Pydew is a range of hills forming the southern edge of Snowdonia National Park. None of them is a significant mountain in its own right, being in general rounded hills. At the north end rises Foel Cwm Sian Llwyd, topped by an ancient burial cairn. To the south is Cyrniau Nod, which seems to compete with Foel Cedig as to which 'summit' is the tallest. Cyrniau Nod's pile of stones appears to be the highest. Wholly within the national park are the three final summits in the vicinity—Pen y Cerrig Duon, Foel y Geifr and Foel Goch. The northern slopes of these hills are afforested as part of Penllyn Forest. Rising east of the Bwlch y Groes is the mountain of Moel y Cerrig Duon, the summit cairn being fairly quickly reached from the minor roads that cross the hills hereabouts.

The Aran hills are an attractive range of mountains, with rocky outcrops and steep corries

54

adding to their attraction. Just off the main range are the lower summits of Esgeiriau Gwynion and Foel Hafod-fynydd. From Llanuwchllyn a long footpath traverses the Arans, the first mountain reached being Aran Benllyn, its eastern cliffs being quite spectacular. Erw y Ddafad-ddu follows, almost as tall, and still having steep eastern cliffs, but the actual summit is fairly flat. The highest mountain follows, Aran Fawddwy, which at 2,969 feet is just short of Welsh Munro status.

Glasgwm is a tall rounded summit, its near-vertical cliffs being lower down to the east. Its summit is at the cairn just north of the little tarn. To its south-west, beyond the forest plantation, is Pen y Brynfforchog, a summit that can be climbed from the Bwlch Oerddrws pass. Off Arenig Fawddwy are two other mountain summits, Gwaun y Llwyni, with its attractive northern corrie, and the long ridge of Pen yr Allt Uchaf.

South of Dollgellau are two ranges of tall hills, the better known Cadair Idris and the lesser known hills south of Bwlch Oerddrws. In the latter group is Maesglase, or Maen Du, a tall and attractive summit when viewed from the valley below. To its west is Cribin Fawr and Waun-oer, both having rocky eastern slopes.

Cadair Idris is a spectacular range of summits, culminating in Penygadair, a significant summit seen from Dolgellau and places to the north. The other two mountains on this horizon are Mynydd Moel and Cyfrwy. To the south of Penygadair is Craig Cwm Amarch (sometimes referred to as Mynydd Pencoed), the summit cairn perched on the edge of the steep cliffs that drop to Llyn Cau in its corrie. Much of Cadair Idris is included in the national nature reserve, and parts of its slopes are owned by the National Trust.

The ridge west of Cadair Idris includes the summits of Craig-las and Craig-y-llyn, both attractive summits in their own right. North of Machynlleth are the two outliers of Tarren y Gesail and Tarrenhendre.

54

Height	Name	NGR	OS L	OS E	Ascent
	Berwyn:				
630	Moel Fferna	SJ 117398	125	E255	
621	Moel Gwyn	SJ 089369	125	E255	
784	Cadair Bronwen	SJ 077346	125	E255	
691	Foel Wen	SJ 099333	125	E255	
681	Mynydd Tarw	SJ 112323	125	E255	
830	⊙ Cadair Berwyn	SJ 072323	125	E255	
827	Moel Sych	SJ 066319	125	E255	
665	Post Gwyn	SJ 048293	125	E255	
648	Foel Cwm Sian Llwyd	SH 997314	125	E255	
667	Cyrniau Nod	SH 989279	125	E239/E255	
646	Pen y Boncyn Trefeilw	SH 962283	125	E23	
626	Foel y Geifr	SH 937275	125	E23	
613	Foel Goch	SH 943291	125	E23	
625	Moel y Cerrig Duon	SH 923241	125	E23	
671	Esgeiriau Gwynion	SH 889236	124/125	E23	

54

Height	Name	NGR	OS L	OS E	Ascent
614	Llechwedd Du	SH 894223	124/125	E23	
885	Aran Benllyn	SH 867242	124/125	E23	
872	Erw y Ddafad-ddu	SH 865234	124/125	E23	
689	Foel Hafod-fynydd	SH 877227	124/125	E23	
905	⊙Aran Fawddwy	SH 862222	124/125	E23	
685	Pen Main (Gwaun y Llwyni)	SH 857205	124/125	E23	
625 est	Pen yr Allt Uchaf	SH 871197	124/125	E23	
779	⊙Craig y Ffynnon (Glasgwm)	SH 837194	124/125	E23	
685	Pen y Brynfforchog	SH 817179	124/125	E23	
674	Maesglase - Maen Du	SH 822151	124/125	E23	
659	Cribin Fawr	SH 796152	124	E23	
670	Waun-oer	SH 786147	124	E23	
	Cader Idris:				
863	Mynydd Moel	SH 728137	124	E23	
893	⊙Cader Idris - Pen y Gadair	SH 712131	124	E23	
811	Cyfrwy (The Saddle)	SH 704134	124	E23	

54

Craig Cwm Amarch	791	SH 710121	124	E23		
Craig-las	661	SH 677135	124	E23		
Craig-y-llyn	622	SH 666119	124	E23		
Tarren y Gesail	667	SH 710059	124	E23		
Tarrenhendre	634	SH 683041	135	E23		

54

Pumlumon & Llanwrthwl

This section includes a random selection of summits in central Wales, located across the Ceredigion/Powys border. The Pumlumon range is located on the north side of the A44, and it is here that both the rivers Severn and Wye have their sources. The source of the Severn lies to the north of Pen Pumlumon Arwystli (the mountain is also known without the 'Pen' part) and the Severn Way path leads to it. The summit of the mountain is located at some ancient burial cairns, a large boundary stone marking the highest point.

Just above the source of the Wye is a mountain named Pen lluest-y-carn, but some accounts name it as Pen Pumlumon Llygad-bychan, a name made up by George Bridge. The summit has a cairn and a boundary stone inscribed 1865.

The highest of the Pumlumon (or Plynlimon) group is Pen Pumlumon Fawr, again topped with Bronze Age cairns. This summit is more dramatic than the others in this range, and it well-deserves its status as the tallest point. An outlier on a ridge stretching south of Pen Pumlumon Fawr is Y Garn, or Drum Peithnant, the summit of which has a prehistoric cairn.

To the south of the A44 the extensive Cefn Croes wind farm extends across the moors and forests. At the southern end of the wind farm rises Pen y Garn, which only just makes the list as a mountain, being around 2,003 feet tall. Its summit has a prehistoric cairn and a trig point.

West of Newbridge on Wye two of the hills rise above 2,000 feet. These are Gorllwyn and Drygarn Fawr, both of which have a prehistoric cairn marking the top,

A section of the Welsh hills at Radnor Forest rise above 2,000 feet. Three rather flat but steep-sided summits qualify as mountains, two of them on opposite sides of Harley Dingle shooting range, the third lying further east, above Kinnerton. It is possible to make a circuit round all three from New Radnor.

Height	Name	NGR	OS L	OS E	Ascent
	Plynlimon:				
741	Pen Pumlumon Arwystli	SN 815877	135	E214	
727	Pen Lluest-y-Cárn	SN 799872	135	E213/E214	
752	Plynlimon (Pen Pumlumon Fawr)	SN 789869	135	E213	
684	Drum Peithnant (Y Garn)	SN 775852	135	E213	
610	Pen y Garn (Bryn Garw)	SN 799771	135/147	E213	
	Llanwrthl Hills:				
645	Drygarn Fawr	SN 863584	147	E187/E200	
612	Gorllwyn	SN 918591	147	E200	
	Radnor Forest:				
660	Great Rhos	SO 182638	148	E200	
650	Black Mixen	SO 196643	148	E200	
610	Bache Hill	SO 214636	137/148	E201	

55

Mynydd Du, Fforest Fawr & Brecon Beacons

All of this section is located within the Brecon Beacons National Park, which was established in 1957 as Britain's tenth national park. The Black Mountain (or Mynydd Du as it is known in Welsh, and which is often used to distinguish the range from the Black Mountains, which straddles the English border) is the westmost part of the group.

There are five hills that qualify as mountains within the Black Mountain, all located between the A4069 and A4067 roads. The westmost reaches 2,022 feet, and is known as Moel Gornach, though sometimes as Garreg Lwyd. A path from the main road soon reaches the huge summit cairn, which had its origins as a burial cairn.

The path over Foel Fraith (which just fails to qualify as a mountain) can be followed to Garreg Las, the summit marked by the ancient Carnau'r Gareg-las, two burial cairns, the northmost of which is the highest point. The ridge is crossed by the Beacons Way long distance path.

To the east are three attractive mountains with steep cliffs on their northern or eastern slopes. Picws Du, or Bannau Sir Gaer, is the westmost, its northern slopes tumbling to Llyn y Fan Fach from the prehistoric summit cairn. Fan Brycheiniog is the tallest summit hereabouts, the trig point and prehistoric cairn Twr y Fan Foel vying for superiority—climbers should visit both! South from here is the long ridge of Fan Hir.

Between the A4067 and the A470, the hills are a bit more retiring, being rounded summits with fewer examples of rocks breaking through the surface. This is the Fforest Fawr range. At the western end, rising steadily above the Bwlch Bryn-rhudd, is Fan Gihirych, a flat-topped summit with an attractive corrie to one side. To the east one can see Fan Nedd, a flattish top with steep sides.

On the east side of the Llethr pass the hills are slightly more interesting, Fan Llia being traversed by the Beacons Way, which continues north-eastwards towards Craig Cerrig-gleisiad. This name actually refers to the steep rocky slopes to the north-east of the summit, and some folk refer to the hill as Rhos Dringarth. Just to the north is Fan Frynych, the summit of which has old mineral workings. These two summits are now part of the Graig Cerrig Gleisiad national nature reserve and are owned by the Nature Conservancy Council. To the south rises Fan Fawr, quickly climbed from the Storey Arms on the A470.

The Brecon Beacons spread east from the Storey Arms pass on the A470 to the Usk valley. The western section of 8,192 acres is owned by the National Trust, gifted to them in 1965 by the Eagle Star Insurance Company. At the western end, Corn Du probably only just qualifies as a separate mountain. It adjoins Pen y Fan, the tallest of the range at 2,907 feet. Like most of the surrounding mountains, it comprises of Old Red Sandstone, the bedding apparent where the rocks break the surface. To the east is Cribyn, another sandstone summit with ridges fanning out in three directions.

The flatness of the hilltops in the section of the Beacons east of Bwlch ar y Fan means that summits such as Fan y Big and Bwlch y Ddwyallt don't qualify as mountains due to their lack of prominence. At the eastern end of this stretch of upland country is Waun Rydd, a flat-topped hill that claims its mountain status simply as a matter of being slightly taller than its surroundings. The actual highest point is at the northern end of the curved summit, the tall Carn Pica located at a slightly lower point.

The final mountain in the Brecon Beacons group is an outlier to the south. Cefn yr Ystrad is 2,024 feet in height, a limestone summit, much disturbed by the quarries on its northern slopes. On the summit is a trig point and two prehistoric cairns, and from here attractive views of the high Beacons are had.

Height	Name	NGR	OS L	OS E	Ascent
	Mynydd Du:				
616	Moel Gornach (Garreg Lwyd)	SN 740179	160	E12	
635	Garreg Las	SN 777202	160	E12	
749	Bannau Sir Gaer (Picws Du)	SN 812218	160	E12	
802	⊙Fan Brycheiniog	SN 825218	160	E12	
761	Fan Hir	SN 831209	160	E12	
	Fforest Fawr:				
725	Fan Gihirych	SN 881191	160	E12	
663	Fan Nedd	SN 913184	160	E12	
632	Fan Llia	SN 938185	160	E12	
734	Fan Fawr	SN 970193	160	E12	
629	Craig Cerrig-gleisiad	SN 960216	160	E12	
629	Fan Frynych	SN 957228	160	E12	
	Brecon Beacons:				
873	Corn Du	SO 007214	160	E12	
886	⊙Pen y Fan	SO 012216	160	E12	

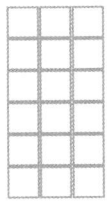

795	☐	Cribin	SO 023213	160	E12
769	☐	⊙Waun Rydd	SO 062207	160	E12
617	☐	Cefn yr Ystrad	SO 087137	160	E12

56

Black Mountains

The Black Mountains, or Mynyddoedd Duon in Welsh, form a series of ridges north of Crickhowell and East of Talgarth, spreading to the English border. The rocks are in the main comprised of Old Red Sandstone, often seen in layers. Black Mountain itself is straddled by the border, and Offa's Dyke Path takes many ramblers over its summit. This is fairly indistinct, Hay Bluff at the northern end being more notable.

All of the summits are located within the Brecon Beacons National Park, which extends west from the border over this section of upland Wales. The northern edge of the mountains forms a great escarpment, the layered rocks being notable points on the skyline from the north. Chwarel y Fan is a summit on one of the ridges that the hills are noted for, a footpath extending over the whole length. Its summit has a large cairn and the remnants of a quarry, from which the hill probably gained its name. Twmpa, also known as Lord Hereford's Knob, is fairly quickly ascended from Gospel Pass, a high moorland road from the Vale of Ewyas to Hay-on-Wye.

Waun Fach is the tallest of the Black Mountains, a large bulk of rounded top often climbed from The Forest to the west. Its summit is little more than a wide peat bog. A wide ridge to the south-east leads to Pen y Gadair Fawr, the summit of which has a prehistoric cairn marking the top. A different ridge can be followed west then south to Mynydd Llysiau,.

The two summits of Pen Allt-mawr and Pen Cerrig-calch have a number of Bronze Age burial cairns either on or near their summits. Pen Cerrig-calch is a significant mountain seen from the village of Crickhowell to the south. Its summit has a number of shake holes across it.

West of the A479, Talgarth to Crickhowell road, rises Mynydd Troed, a substantial elongated summit that just fails to reach the magical 2,000 foot mark, coming in at 1,998 feet.

Height	Name	NGR	OS L	OS E	Ascent
703	Black Mountain (Moel Olchon)	SO 256350	161	E13	
690	Twmpa (Lord Hereford's Knob)	SO 225350	161	E13	
679	Chwarel y Fan	SO 258294	161	E13	
811	⊙Waun Fach	SO 215299	161	E13	
800	Pen y Gader Fawr	SO 229287	161	E13	
663	Mynydd Llysiau	SO 207279	161	E13	
719	Pen Allt-mawr	SO 207243	161	E13	
701	Pen Cerrig-calch	SO 217223	161	E13	

Dartmoor

The extensive upland area of Dartmoor is located in Devon and forms one of England's national parks, created in 1951. This is located south of the busy A30, north of the A38 and east of the A836. The secondary roads through the middle part of the moor, where the West Dart River has cut a valley, divides the moor into two, and it is the northern part where the highest hills are located.

The tallest summit in Dartmoor is known as High Willhays, rising above the town of Okehampton. It is not the most well-known summit in the district, however, for the adjoining Yest Tor is better known, being that summit visible from Okehampton, and which is more rugged in appearance. However, those who climb one of these summits will invariably climb the other.

Yest Tor is 2,031 feet in height, a separate 2,000 foot contour indicating that it is a subsidiary top. Historically, it was thought that Yest Tor was the higher of the two tops, but accurate measurement has shown High Willhays to be eight feet taller. The summit is crowned with a prehistoric cairn and an Ordnance Survey trig point. A path and track from Okehampton, passing the Okehampton Camp, make their way to the top. This camp is a base occupied by the British Army, for Okehampton Common is used for training and red flags are flown when live shooting is taking place. High Willhays and Yest Tor are located within the Danger Area indicated on the maps.

The hills hereabouts are composed of granite, and many rocky tors and other outcrops exist. Although only the two points of the moor exceed 2,000 feet, there are a few other hills that are in excess of 600 metres—Hangingstone Hill and Cut Hill, both of which are 603m above sea level. Both summits are fairly flat, though Hangingstone Hill is topped by a prehistoric burial cairn. Both summits also fall within the danger area.

58

Height	Name	NGR	OS L	OS E	Ascent
☐ 621	High Willhays	SX 580892	161	E28	☐☐☐☐☐☐

Isle of Man

The Isle of Man, located in the midst of the Irish Sea, is a Crown Dependency of Great Britain. The Ordnance Survey compiles maps of the island, hence its inclusion within this guide, although they have not published an Explorer map. There is only one summit on the island that exceeds 2,000 feet—that of Snaefell. The summit is located fairly centrally on the island, though slightly north of centre, and rises steeply above Laxey to the east. Early explorers claimed the summit to be only 1,740 feet or so above sea level, the first to note it being over 2,000 feet being J. F. Berger in 1814. It was not until 1868 when the Ordnance Survey carried out their surveys that the present height of 2,034 feet was confirmed.

The name Snaefell comes from the old Norse for 'snow mountain'. The summit is easily reached by walking from the Bungalow Railway Station on the A18 road. From there a rough path and track makes its way to the summit, a distance of around one mile. A couple of transmitting masts and associated buildings are located near the top, which is marked by an Ordnance Survey trig point.

There is another method of reaching the top of Snaefell—by taking the Snaefell Mountain Railway, an electric train that makes its way to the top from the village of Laxey. It was established in 1895 and only operates in the summer months. The average incline is 1 in 12 and the railway gauge is 3.5 feet. Associated with the railway is a summit café, formerly the Summit Hotel

Traditionally, it is said that six kingdoms can be seen from the summit on a clear day—the kingdoms of Man, England, Ireland, Wales, Scotland, and Heaven. Certainly, the view is extensive, with the Mountains of Mourne visible in Ireland, the Lake District in England, Snowdonia in Wales and Galloway in Scotland possible on clear days. An indicator on the summit points the out the various landmarks.

59

Height	Name	NGR	OS L	OS E	Ascent
☐ 621	Snaefell	SC 398881	95	N/A	☐☐☐☐☐☐

Bibliography

Bridge, George, *The Mountains of England and Wales*, Gaston's Alpine Books/West Col Productions, Goring, 1973.

Corbett, J. Rooke, *Twenty-Fives*, Rucksack Club Journal, 1929.

Dawson, Alan, *The Relative Hills of Britain,* Cicerone Press, Milnthorpe, 1992.

Docharty, W. MacKnight, *A Selection of Some 900 British and Irish Mountain Tops*, Privately Published, 1954.

Elmslie, Rev W. T., *The Two Thousand Footers of England*, Fell & Rock Climbing Club Journal, 1933.

Fettes, Paul, & Friends, *The Archies: Scotland's 1,000 Metre Mountains*, The Archie Foundation, Dundee, 2017.

Marsh, Terry, *The Summits of Snowdonia: a Guide to the 600-metre Mountains of Snowdonia*, Robert Hale, London, 1984.

Moss, E., *All Those Two-Thousands*, Rucksack Club Journal, 1952.

Munro, Sir Hugh, *et. al, Munro's Tables and Other Tables of Lesser Heights,* Scottish Mountaineering Trust, Edinburgh, 1974.

Nuttall, John and Anne, *The Mountains of England and Wales, Volume 1: Wales*, Cicerone Press, Milnthorpe, 1989.

Nuttall, John and Anne, *The Mountains of England and Wales, Volume 2: England*, Cicerone Press, Milnthorpe, 1989.

Acknowledgments

The compilation of these tables commenced when I was still at school, and my geography teachers allowed access to their map drawers. I can still picture the shape of Cranstackie on the flat One Inch map that started the process.

Ken Wilson of Diadem publishing was very keen to produce the book in the early days, but just couldn't get the process done, and a second publisher was closed by the bank a week after confirming that they would produce the book! The late Irvine Butterfield was a great advocate of the tables, and promoted them in his *High Mountains of Britain and Ireland* book.

The cover illustration of Suilven, drawn by Gillian Love, is based on an image by John MacPherson (www.john-macpherson-photography.com).

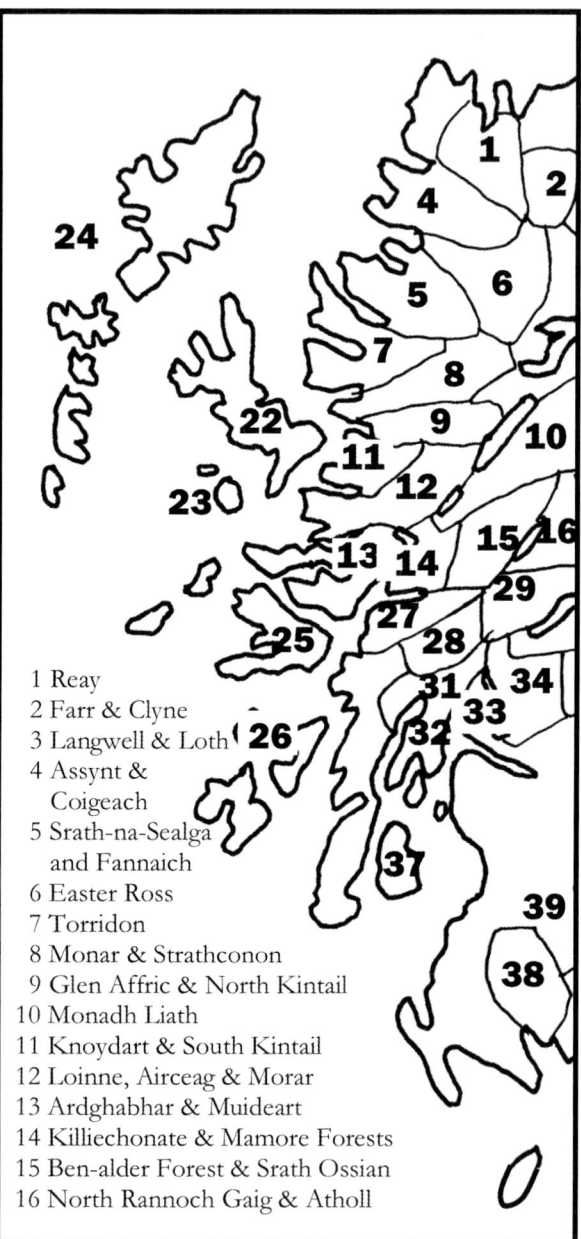

1 Reay
2 Farr & Clyne
3 Langwell & Loth
4 Assynt &
  Coigeach
5 Srath-na-Sealga
  and Fannaich
6 Easter Ross
7 Torridon
8 Monar & Strathconon
9 Glen Affric & North Kintail
10 Monadh Liath
11 Knoydart & South Kintail
12 Loinne, Airceag & Morar
13 Ardghabhar & Muideart
14 Killiechonate & Mamore Forests
15 Ben-alder Forest & Srath Ossian
16 North Rannoch Gaig & Atholl

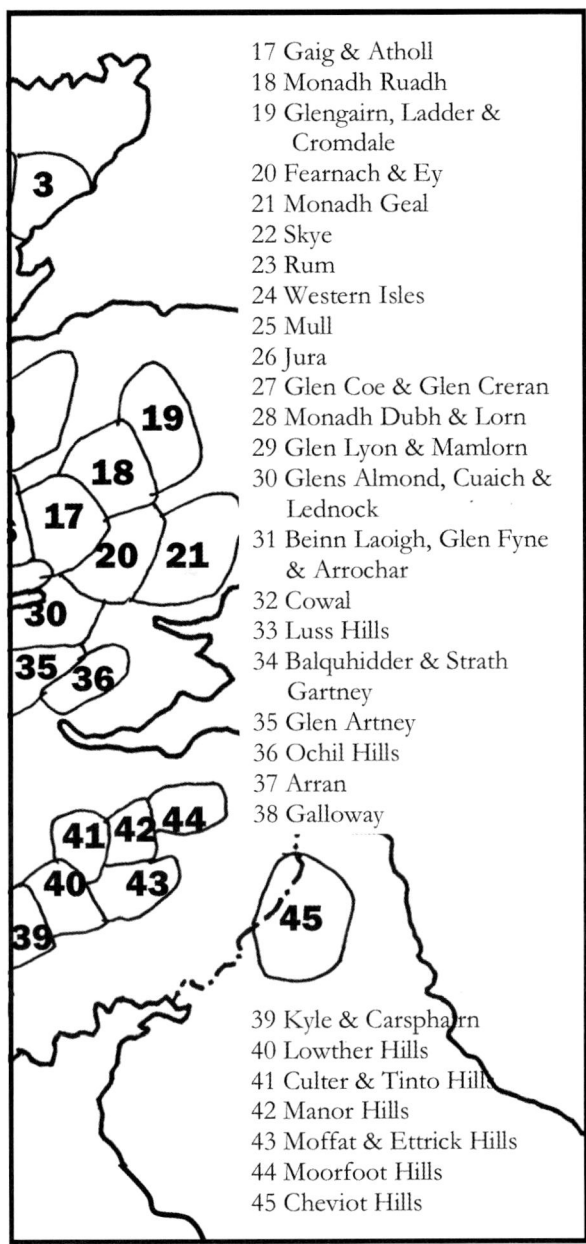

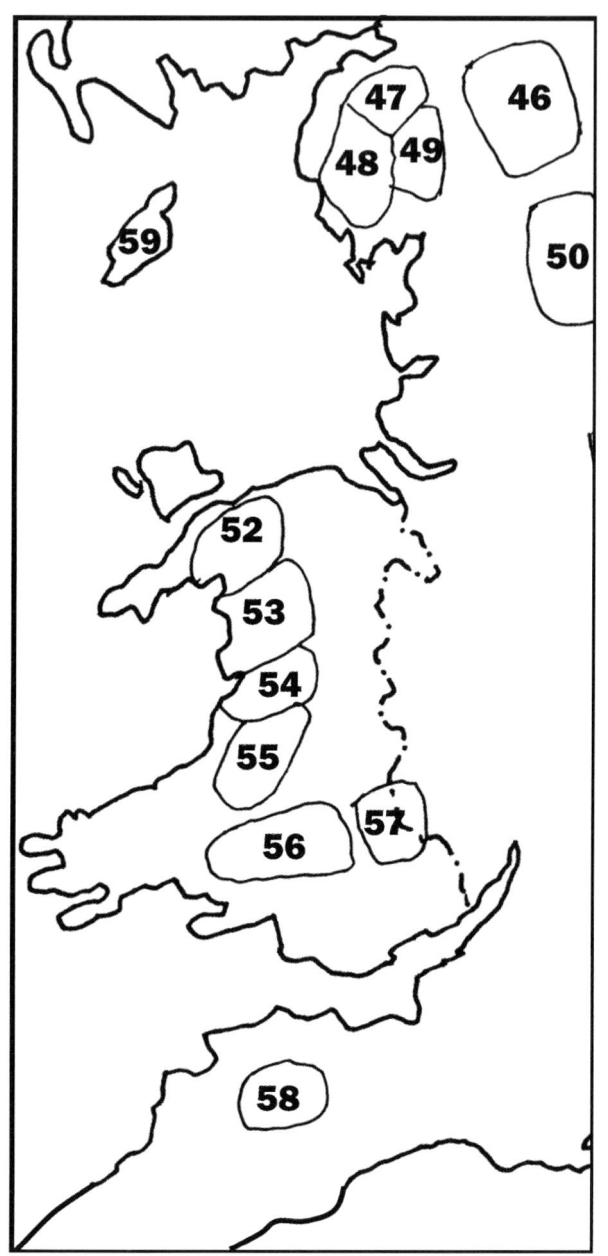

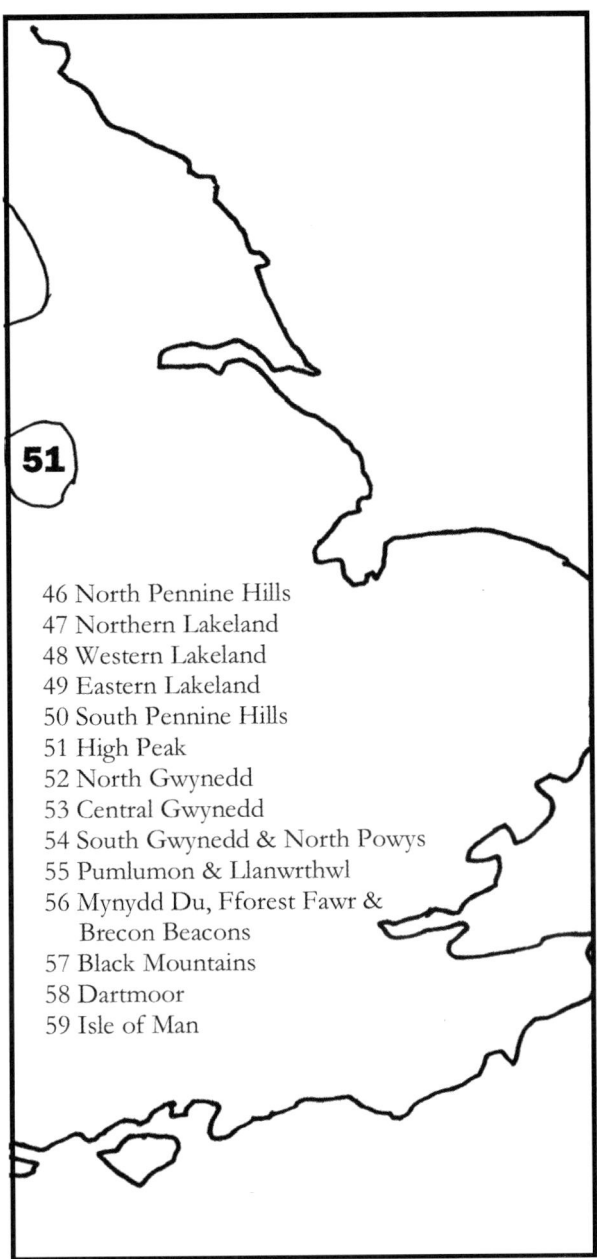

51